Sascha Trippe
TEN THOUSAND STARS AND ONE BLACK HOLE

Sascha Trippe

Ten Thousand Stars and One Black Hole

A Study of the Galactic Center in the Near Infrared

Bibliografische Information Der Deutschen Bibliothek
Die Deutsche Bibliothek verzeichnet diese Publikation in der Deutschen Nationalbibliografie; detaillierte bibliographische Daten sind im Internet über http://dnb.ddb.de abrufbar.

Dissertation der Fakultät für Physik der Ludwig-Maximilians-Universität München zur Erlangung des Grades Doktor der Naturwissenschaften (Dr. rer. nat.)

erstellt am Max-Planck-Institut für extraterrestrische Physik, Garching

vorgelegt von Sascha Trippe, geboren am 13.10.1978 in Düsseldorf

München, den 30. Januar 2008

1. Gutachter: Prof. Dr. Reinhard Genzel
2. Gutachter: Prof. Dr. Ortwin Gerhard

Datum der mündlichen Prüfung: 13. März 2008

©HARLAND media, Lichtenberg (Odw.) 2008
www.harland-media.de

Gedruckt auf alterungsbeständigem Papier nach ISO 9706 (säure-, holz- und chlorfrei).

Printed in Germany

ISBN 978-3-938363-22-5

Contents

List of Figures   X

List of Tables   XI

Zusammenfassung   XIII

Summary   XV

1 Introduction   1
    1.1 The Galactic Center 1
       1.1.1 Scales and structures 2
       1.1.2 The central star cluster 8
       1.1.3 Sagittarius A* 15
    1.2 Observations 21
       1.2.1 Obtaining diffraction-limited data 21
       1.2.2 Instruments 24
    1.3 Data processing and analysis 29
       1.3.1 Data reduction 31
       1.3.2 Deconvolution 35
       1.3.3 Fluxes and positions 39
       1.3.4 Time series 42

2 The LBV candidate GCIRS34W   47
    2.1 Introduction 47
    2.2 Observations 49
       2.2.1 Photometry 49
       2.2.2 Spectroscopy 49
    2.3 Results and discussion 50
    2.4 Summary 60

3 Polarised emission from SgrA* — 61
- 3.1 Introduction — 61
- 3.2 Observations and data reduction — 62
- 3.3 Results — 67
- 3.4 Context — 70
- 3.5 Discussion — 77
  - 3.5.1 Nature of the flares — 77
  - 3.5.2 Geometry of the system — 79
  - 3.5.3 Proposed model — 80
- 3.6 Summary — 81

4 Motion of the stellar cusp — 83
- 4.1 Introduction — 83
- 4.2 Radio Observations — 84
  - 4.2.1 VLA Observations — 85
  - 4.2.2 SiO Maser Proper Motions — 88
- 4.3 Infrared Observations — 92
- 4.4 Radio & Infrared Frame Alignments — 94
- 4.5 Enclosed Mass versus Radius from SgrA* — 95
  - 4.5.1 Dark Matter in the Central Stellar Cluster — 97
  - 4.5.2 IRS 9 not bound to the central parsec — 97
  - 4.5.3 R_0 exceeds 9 kpc — 100
  - 4.5.4 Non-zero V_{LSR} for SgrA* — 101
  - 4.5.5 IRS 9 is (or was) in a binary — 101
- 4.6 Conclusions — 102

5 Kinematics of the CO star cluster — 104
- 5.1 Introduction — 105
- 5.2 Observations and data reduction — 107
- 5.3 Astrometry — 108
  - 5.3.1 The procedure — 108
  - 5.3.2 Geometric distortion — 109
  - 5.3.3 Image registration and mosaicking — 112
  - 5.3.4 Positions and coordinates — 114
  - 5.3.5 Proper motions — 115
- 5.4 Spectroscopy — 117
- 5.5 Results and discussion — 119
  - 5.5.1 Global rotation — 119

	5.5.2	Phase-space distributions	123
	5.5.3	Distribution of stellar 3D speeds	129
	5.5.4	Statistical parallax of the Galactic Center	131
	5.5.5	Acceleration upper limits	135
	5.5.6	The star group IRS13E	137
5.6	Conclusions		140

6 Conclusions   **142**

Acknowledgements   **146**

Bibliography   **147**

List of Figures

1.1	Optical image of the Galactic Center as seen from Cerro Paranal, Chile, in July 2007	2
1.2	90 cm wide-field radio map of the Galactic Center	3
1.3	Radio map of the mini-spiral	4
1.4	Radio map of the circumnuclear disk	5
1.5	Near infrared colour map of the central parsec of the Milky Way	6
1.6	The innermost arcseconds of the Milky Way	7
1.7	An early infrared map of the central star cluster	10
1.8	K-band imge of the central star cluster	11
1.9	Surface density of stars and fraction of late type stars vs. projected radius from SgrA*	12
1.10	K-band luminosity functions of the central cluster	13
1.11	Important planar structures in the GC	14
1.12	Enclosed mass vs. projected distance from SgrA*	16
1.13	Stellar orbits around SgrA*	17
1.14	K-band flare from SgrA* as observed in May 2006	18
1.15	Lightcurve of a bright L-band flare obtained in April 2007	19
1.16	Spectral energy distribution of emission from SgrA*	20
1.17	Speckle imaging observation of a binary star	22
1.18	Sketch of an adaptive optics system	23
1.19	Optical system of the speckle imaging camera SHARP I	25
1.20	Principles of integral field spectroscopy	26
1.21	The integral field spectrometer 3D.	27
1.22	Overview images of the Very Large Telescope	28
1.23	Optical design of the camera system NAOS/CONICA	29
1.24	View into the integral field spectrometer SPIFFI	30
1.25	An illustration of the necessity of data reduction	31
1.26	Wavelength calibration image used for reducing SPIFFI data	34
1.27	Illustration of different deconvolution techniques	37

List of Figures

1.28	Analyzing a spectral line by fitting an analytical profile to it.	39
1.29	Measuring the radial velocity of a CO absorption line star via template correlation.	41
1.30	Time series analysis with a Scargle periodogram.	43
1.31	Period backfolding analysis of a time series.	44
2.1	H and K band lightcurves and colours of IRS34W	52
2.2	Normalized K band spectra of IRS34W	54
2.3	Approximated position of IRS34W in a HRD compared to already known similar stars	59
3.1	Four examples for the calculation of degrees and angles of polarisation	63
3.2	Comparison of two methods for calculating polarisation parameters	64
3.3	Observed total and polarised fluxes of SgrA* and the comparison stars S2 and S7	65
3.4	Contour maps of integrated and polarised emission from SgrA*.	66
3.5	Evolution of degree and angle of polarisation for SgrA* and S2 during the flare	68
3.6	A beginning H band flare observed on April 28, 2004, with NACO	71
3.7	Lightcurves and power spectra illustrating the short-time sub-structure of SgrA* flares	75
3.8	An artificial flare lightcurve as illustration of sub-structure signatures.	76
4.1	Composite spectrum of stellar SiO masers detected with the VLA in 1998, 2000, and 2006	88
4.2	Position offsets from SgrA* versus time for the eight SiO maser stars with negative LSR velocities	89
4.3	Position offsets from SgrA* versus time for the seven SiO maser stars with positive LSR velocities	90
4.4	Infrared (K-band) image of the central ±20" of the Galactic Center	91
4.5	Constraints on the enclosed mass as a function of radius from SgrA*.	98
5.1	K-band mosaic of the Galactic Center cluster obtained in March 2007.	106
5.2	Residual image alignment errors before and after distortion correction and image registration.	111
5.3	Statistical uncertainty of absolute radio reference frame coordinates as a function of position.	113
5.4	Histograms of proper motion errors.	116
5.5	Histogram of statistical errors in line-of-sight velocity.	117

5.6	K magnitude distributions of all proper motion stars and 3D velocity stars.	118
5.7	Modulations in stellar proper motions.	119
5.8	Signatures of rotation in proper motions and radial velocities.	121
5.9	Histograms of proper motions in l and b of all proper motion stars.	122
5.10	Velocity-coordinate phase-space maps.	124
5.11	Velocity-velocity phase-space maps.	125
5.12	4D two-point correlation functions for all proper motion stars.	126
5.13	Distribution of stellar speeds for all 3D velocity stars.	130
5.14	Histograms of proper motions and of radial velocities.	132
5.15	Acceleration upper limits vs. projected distance from SgrA*.	134
5.16	Ratio of minimum physical distance and projected distance from SgrA* vs. projected distance.	136
5.17	Proper motion maps for the IRS13E group.	138

List of Tables

2.1	H and K magnitudes and colours of IRS34W	53
2.2	Equivalent widths of the dominant emission lines in the March 1996, April 2003 and August 2004 spectra	55
3.1	Properties of infrared flares observed since 2002	72
4.1	SiO maser astrometry	87
4.2	SiO maser proper motions	92
4.3	3-dimensional stellar motions and enclosed mass limits	93
4.4	Radio–infrared proper motion alignment	95
5.1	Properties of three stars with speeds $v_{3D} > 438$ km/s.	131

Pervigilans noctem carpit.

— *origin unknown*

Zusammenfassung

Gegenstand dieser Dissertation sind die physikalischen Eigenschaften der Zentralregion – genauer: der innersten $\sim$2 pc – unserer Milchstrasse. Ziel war es hierbei, neue Erkenntnisse über Struktur, Zusammensetzung und Dynamik des zentralen Sternhaufens, einschliesslich des supermassiven ($\sim 4 \cdot 10^6 \, M_\odot$) Schwarzen Lochs Sagittarius A* (SgrA*), zu gewinnen.

Diese Arbeit beruht auf Bild- und integralfeld-spektroskopischen Daten, die mit den Instrumenten NAOS/CONICA und SINFONI am Very Large Telescope der Europäischen Südsternwarte gesammelt wurden. Diese Daten decken den Spektralbereich des nahen Infrarotlichtes (NIR) in der Region $\sim$1.5-4μm (H- bis L-Band) ab. In Einzelfällen wurden zusätzlich Infrarot- und Radiodaten anderer Observatorien einbezogen.

Ein auffälliges Mitglied des Sternhaufens ist der Ofpe/WN9-Stern GCIRS 34W. Dieser Stern zeigt eine starke ($\Delta K \simeq 1.5$), unregelmäßige Leuchtkraftvariabilität auf Zeitskalen von Monaten bis Jahren; daher wurde dieses Objekt als candidate Luminous Blue Variable (cLBV) eingestuft. Um die dahinterstehenden Mechanismen zu verstehen, haben wir photometrische und spektroskopische Daten, die einen Zeitraum von knapp 13 Jahren abdecken, ausgewertet. Aus diesen Daten konnten wir schließen, dass die photometrische Veränderlichkeit durch eine Staubhülle mit variabler Struktur hervorgerufen wird. Dieser Staub wird offenbar durch den Stern in unregelmäßigen Ausbrüchen freigesetzt, in Übereinstimmung mit dem LBV-Szenario.

SgrA* zeigt im NIR gelegentliche ($\sim$3-4 pro Tag) Helligkeitsausbrüche ("flares"), die jeweils etwa 1-2 Stunden andauern. Eine dritte charakteristische Zeitskala von $\sim$15-20 Minuten ist durch eine gelegentlich beobachtete quasiperiodische Modulation der flare-Lichtkurven gegeben. Diese flares, und insbesondere die drei charakteristischen Zeitskalen, sind bisher nicht gut verstanden. Am 30. Mai 2006 gelang uns die polarimetrische Beobachtung eines besonders hellen flares. Dieser zeigte starke (bis $\sim$40%) lineare Polarisation. Integrierter Fluss, polarisierter Fluss und Polarisationswinkel waren zeitlich variabel auf einer 15-min-Skala. Diese Beobachtungen diskutieren wir in Zusammenhang mit früheren Ergebnissen aus Wellenlängenbereichen von

Radio bis Röntgen. Wir konnten schliessen, dass sich unsere Beobachtungen in guter Übereinstimmung mit einem dynamischen Emissionsmodell befinden. In diesem Modell geht die Emission von einer Plasmablase aus, die sich in der Akkretionsscheibe um SgrA* durch magnetische Rekonnexion oder Einfall von Materie bildet. Diese Blase bewegt sich auf einem Orbit im Bereich der innersten stabilen kreisförmigen Umlaufbahn (entsprechend einer Periode von ~15-20 min) um die Zentralmasse. Dabei wird sie auf einer Zeitskala von etwa einer Stunde entlang des Orbits geschert, kühlt aus und erlischt.

Der zentrale Sternhaufen zeigt eine komplexe Dynamik bis hin zu Keplerschen Sternorbits um SgrA*. Die kinematische Beschreibung des Haufens erfordert die Bestimmung von stellaren Positionen und Eigenbewegungen mit bestmöglicher Genauigkeit. Im NIR wird die astrometrische Analyse durch das Fehlen einer absoluten Koordinatenreferenz im Galaktischen Zentrum erschwert. Einen Ausweg bietet eine Gruppe von neun SiO-Masersternen, die sowohl im NIR als auch im Radiolicht beobachbar sind. Für diese Sterne sind Radiopositionen relativ zu SgrA* bekannt. Durch Vermessung von Positionen und Bewegungen sowohl im NIR als auch im Radiolicht lassen sich diese beiden Koordinatensysteme ineinander überführen. Um ein Maß für die Stabilität dieser Koordinaten zu erhalten, haben wir die relative Bewegung der beiden Bezugssysteme (Radio vs. NIR) zueinander vermessen. Hierbei ergab sich eine überraschend hohe Relativgeschwindigkeit von 30 ± 10 km/s. Daraus schließen wir, dass die systematische Ungenauigkeit dieser kombinierten Koordinaten größer ist als erwartet und dieses Bezugssystem mit entsprechender Vorsicht verwendet werden muss.

Um die Kinematik des zentralen Sternhaufens bis etwa 1 pc von SgrA* umfassend zu beschreiben, haben wir Eigenbewegungen für etwa 5500 Sterne und Radialgeschwindigkeiten für ca. 660 Sterne gewonnen. Mit Genauigkeiten von 5–10 km/s pro Koordinate ist unsere Untersuchung die mit Abstand umfassendste und präziseste Analyse dieser Art im Galaktischen Zentrum. Wir fanden eine globale Rotation des Sternhaufens im Sinne der Galaktischen Rotation. Die stellaren 3D-Geschwindigkeiten folgen einer Maxwell-Verteilung, was dem Verhalten eines dynamisch relaxierten Systems entspricht. Die 4D-Phasenraumverteilung der Sterne stimmt, ausgedrückt in einer Zweipunkt-Korrelationsfunktion, mit der eines isotropen Rotators auf 2% überein; insbesondere finden sich keine Anzeichen für Sternströme. Aus der 3D-Geschwindigkeitsdispersion ergibt sich eine statistische Parallaxe von 8.37 ± 0.29 kpc. Insgesamt stellt sich der zentrale Sternhaufen als isotropes, dynamisch relaxiertes und phasendurchmischtes System dar.

Summary

This dissertation treats the physical properties of the central region – here meaning the central $\sim$2 pc – of our Milky Way. It aims at understanding of structure, composition, and dynamics of the central star cluster, including the supermassive ($\sim 4 \cdot 10^6\ M_\odot$) black hole Sagittarius A* (SgrA*).

This thesis is based on imaging and integral-field spectroscopy data obtained with the instruments NAOS/CONICA and SINFONI at the Very Large Telescope of the European Southern Observatory. These observations cover the near-infrared (NIR) wavelength regime in the range $\sim$1.5-4μm (H to L bands). In a few cases infrared and radio data collected at other observatories were included into our analysis additionally.

One outstanding cluster member is the Ofpe/WN9 star GCIRS 34W. This source shows strong ($\Delta K \simeq 1.5$) irregular photometric variability on timescales from months to years. Therefore this star had been classified as candidate Luminous Blue Variable (cLBV). In order to understand the responsible physical mechanisms, we analyzed photometric and spectroscopic data covering a timeline of about 13 years. From the combined information we conclude that the photometric variability is caused by a dust shell with a variable structure. This dust is apparently emitted in irregular outbursts of the star, in good agreement with the LBV scenario.

In NIR, SgrA* shows occasional ($\sim$3-4 per day) outbursts ("flares") of radiation lasting about 1-2 hours each. An additional, third characteristic timescale of $\sim$15-20 minutes is defined by a quasi-periodic modulation of flare lightcurves which is sometimes observed. The flares, and especially their three characteristic timescales, are not well-understood. At 30 May 2006 we managed to observed an exceptionally luminous event in polarimetric imaging mode. This flare showed a strong (up to $\sim$40%) linear polarisation. Integrated flux, polarized flux, and polarisation angle showed variations on a 15-min timescale. We discuss these observations in the context of results obtained earlier in wavelengths from radio to X-ray. We were able to conclude that our observations are in good agreement with a dynamical emission model. In this picture, the observed radiation is emitted by a plasma bubble which arises in the accretion disk of SgrA* due to magnetic reconnection or the infall of matter. This bubble orbits the central mass at or close to the innermost stable circular orbit (corresponding to a period of ∼15-20 min). While doing so, it is sheared along its orbit, cools down, and eventually vanishes after roughly one hour.

The central star cluster shows complex dynamics including Keplerian star orbits around SgrA*. A kinematic description of the cluster requires accurate measurements of stellar positions and proper motions. In NIR, this analysis is complicated by the fact that no absolute coordinate reference system is available in the Galactic Center. A loophole is offered by a group of nine SiO maser stars observable in NIR and radio. For these stars radio positions relative to SgrA* are known. By measuring positions and motions in NIR and radio independently, the two reference frames can be connected. In order to check the stability of this coordinate system, we measured the relative motion of the two (radio vs. NIR) reference frames. Surprisingly, the relative velocity turned out to be quite high, about 30 ± 10 km/s. From this we conclude that the systematic uncertainty of the combined coordinate system is larger than expected; thus this reference frame has to be used with appropriate care.

For a detailed kinematic description of the central cluster (up to about 1 pc from SgrA*) we extracted proper motions for about 5500 stars and line-of-sight velocities for about 660 stars. With uncertainties of 5–10 km/s per coordinate, our analysis is the so-far most accurate and comprehensive work of this type for the case of the Galactic Center. We found global cluster rotation in the sense of general Galactic rotation. The 4D phase space distribution of the cluster stars, expressed via a two-point correlation function, agrees with that of an isotropic rotator within 2%; there is no indication for star streams. The stellar 3D speeds follow a Maxwellian distribution, which is in agreement with dynamical relaxation. From the 3D velocity dispersion we obtain a statistical parallax of 8.37 ± 0.29 kpc. In total, the central star cluster appears as an isotropic, uniform, dynamically relaxed, phase-mixed system.

Chapter 1

Introduction

1.1 The Galactic Center

The glowing band of the Milky Way crossing the night sky has been observed and admired by human beings since ancient times. The fact that this band is indeed a large ensemble of individual stars was realized first by Galileo Galilei around 1610. In the middle of the 18th century, Immanuel Kant and Thomas Wright proposed that the Milky Way is a disk of stars and that our sun is located within this system.

A first attempt to map the geometry of the galaxy was made by William Herschel in the 1780s. Based on a star count method, he concluded that the solar system is located close to the geometric center of the Milky Way. Using a more elaborate star count method than Herschel, Jacobus Kapteyn postulated the "Kapteyn universe" model for the description of the Galactic geometry around 1900. Kapteyn found a radius of $\sim$8500 parsecs (pc) for the Milky Way and $\sim$650 pc for the distance of the sun from its center. About 15 years later, Harlow Shapley used the distribution and distances of globular clusters to derive a Galactic radius of $\sim$100 kpc and a Galactic Center (GC) distance of $\sim$13 kpc (Carroll & Ostlie 1996; Reid 1993). For comparison, the modern distance value is 8.0$\pm$0.5 kpc (Reid 1993; the ongoing research will be discussed later in this section).

The central region of the Milky Way, located in the Sagittarius constellation (Oort & Rougoor 1960; see Fig. 1.1), is not observable at optical wavelengths due to strong interstellar extinction of $A_V \simeq 30$. Indeed, first observations of the GC were carried out in the radio and infrared wavelength regimes. In both, radio (Downes & Maxwell 1966) and infrared bands (Becklin & Neugebauer 1968), an increase in light intensity towards the GC was observed. Finally, the geometric center of the Milky Way was identified with the point-like radio source Sagittarius A* (SgrA*) discovered by Balick & Brown (1974).

Figure 1.1: Optical image of the Galactic Center as seen from Cerro Paranal, Chile, in July 2007. The geometric center of the Milky Way is marked by the end of the laser beam, which was emitted by the VLT laser guide star facility while the GC was observed with adaptive optics instruments. *Image: Yuri Beletsky/ESO/MPE.*

1.1.1 Scales and structures

Since the 1950s, the increasing precision with which observations in non-optical wavelength regimes could be carried out has led to a vast amount of observing programmes examining the physical properties of the center of the Milky Way. Observations were carried out from radio to TeV bands. These observations have revealed complex structure within the Galactic Center region. Before discussing individual objects or features, it is useful to sort the phenomena according to they size scales on which they occur.

On spatial scales of a **few 100 parsecs** the GC region is dominated by extended gas emission obvious especially in the radio. Along the Galactic plane, a number of gas filaments and supernova remnants with spatial sizes in the order $\sim 10...100$ pc are located (e.g. Kassim et al. 1999; see Fig. 1.2). Additionally, this area hosts three

1.1. The Galactic Center

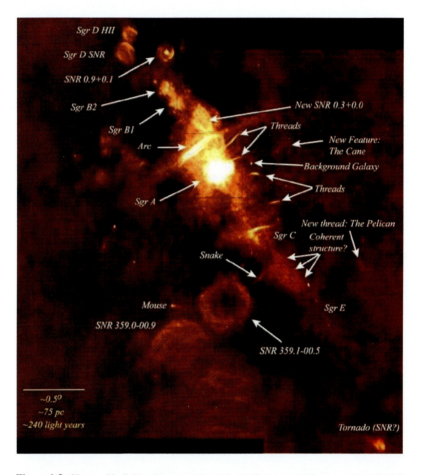

Figure 1.2: 90 cm wide-field radio map of the Galactic Center; north is up, east is to the left. This image gives an overview of the complex large scale structure within the central ∼500 pc of the Milky Way. The region is dominated by supernova remnants (features labeled "SNR", "Sgr" [except SgrA]), the SgrA region contains the geometric center of our galaxy. *Image: Kassim et al. (1999)/NRAO.*

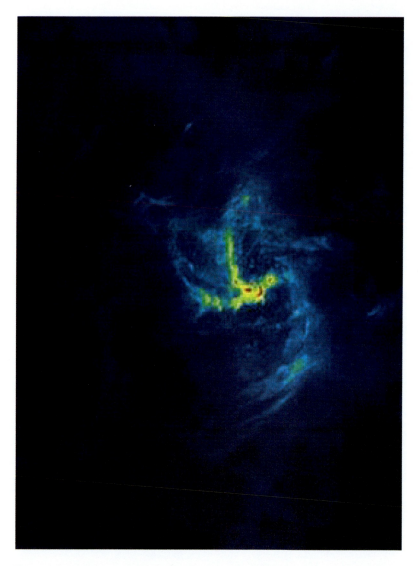

Figure 1.3: Radio map of the central ~3×4 pc. For obvious reasons, the central feature is named mini-spiral. This structure is composed of partially ionized gas, mainly hydrogen. *Image: University of Illinois, Urbana / K.Y. Lo.*

1.1. The Galactic Center

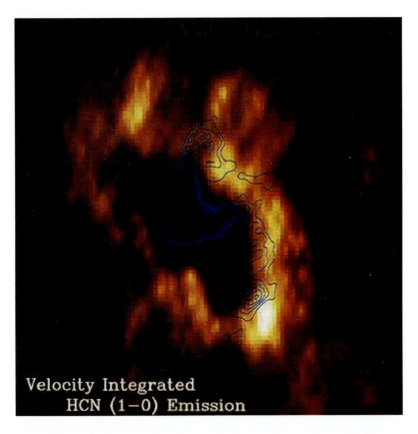

Figure 1.4: HCN molecular radio emission map of the central ∼6×6 pc (yellow/orange). This structure is known as the Circumnuclear Disk. The blue contours show the mini-spiral for comparison. *Image: The University of Arizona / Steward Observatory.*

Figure 1.5: Near infrared colour map of the central parsec (corresponding to ∼25") of the Milky Way. This image is a composite of three maps obtained in H, K, and L bands (1.7–3.8 μm); north is up, east to the left. The innermost region of our galaxy is dominated by a densely populated star cluster pervaded by gaseous filaments.

1.1. The Galactic Center

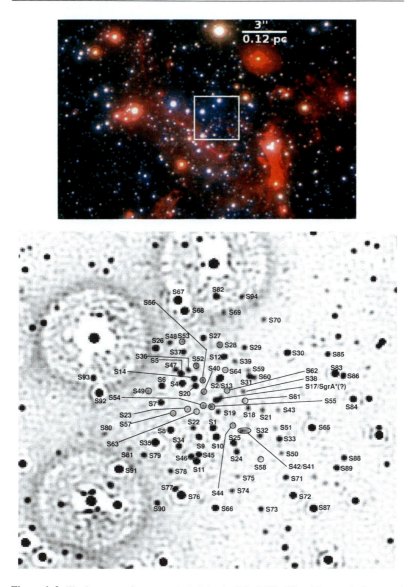

Figure 1.6: The innermost 3 arcseconds (∼0.1 pc) of the Milky Way. *Top panel*: the central part of Fig. 1.5. The white box marks the area shown in the bottom panel. *Bottom panel*: deconvolved H band (1.7 µm) image of the stellar cusp around the Galactic supermassive black hole Sagittarius A* obtained in April 2007. At the epoch of this observation the star S17 (image center) was located at the projected position of SgrA*. *Image: S. Gillessen/MPE.*

massive star clusters located within 50 pc (in projection) from SgrA*: the Arches, Quintuplet, and central (where SgrA* is located in) clusters. These three objects are the most massive ($\sim 10^{4...5} M_\odot$) star clusters in the Milky Way and contain significant populations of young (few Myrs) stars (e.g. Figer et al. 1999).

The central cluster (Fig. 1.5) spans a spatial scale of a **few parsecs** about SgrA*. This cluster is a system composed of two main star populations. The majority population is formed by isotropically distributed old (Gyrs) CO absorption stars. The second population is a small ($\sim$100 objects) group of young (few Myrs) hot H/He emission line stars located in two disks centered on SgrA* (e.g. Krabbe et al. 1991; Genzel et al. 2003a; Paumard et al. 2006). The star cluster is pervaded by partially ionised gaseous (mainly H II) filaments forming a roughly spiral-shaped configuration, thus this structure is named "mini-spiral". In the image plane, the mini-spiral spans over an area of about 1×2.5 pc (Ekers et al. 1983). Slightly further outwards ($\sim$1.5–7 pc from SgrA*) the cluster is surrounded by an additional gaseous structure, the circumnuclear disk (CND; Becklin, Gatley & Werner 1982). Figs. 1.3, 1.4 show mini-spiral and CND in detail.

The **innermost $\sim$0.05 parsecs** of the central cluster (Fig. 1.6) are occupied by a cusp composed of mainly B main sequence stars (the "S-stars"). In this region the proximity to the central black hole leads to accelerations of the stellar trajectories observable within a few years (Schödel et al. 2002; Ghez et al. 2003, 2005a; Eisenhauer et al. 2005a).

The smallest spatial scales – of the order $\sim$**0.1 AU** – for the Galactic Center are given by the Schwarzschild radius of the supermassive ($\sim 4 \cdot 10^6 M_\odot$) black hole SgrA* which is $\sim$0.07 AU (or $\sim 10^{10}$ meters), corresponding to ~ 10 µas in image scale. The region within a few Schwarzschild radii about SgrA* is the source of radiation detectable in wavelength bands from radio to X-rays (see section 3.1 for an overview).

In this dissertation, two of these scales are explored in detail: (a) the central cluster and selected member stars (chapters 2, 4, and 5) and (b) the emission from SgrA* (chapter 3).

1.1.2 The central star cluster

Detailed studies of the central star cluster were made possible when sufficiently (few arcsec) spatially resolved infrared maps of the GC became available. The first map, with an angular resolution of $\sim$2.5", was obtained by Becklin & Neugebauer (1975). This image is presented in Fig. 1.7 and shows already a complex distribution of extended and point sources. More of historical interest is the fact that Becklin & Neuge-

1.1. The Galactic Center

bauer (1975) introduced the IRS[1] numbering system for Galactic Center sources, which is still in use today.

State-of-the-art observations obtained at 8-m-class telescopes (like the VLT, Keck, Subaru, or Gemini observatories) achieve a diffraction-limited angular resolution of $\sim$60 mas in K-band (2.2 µm) when using adaptive optics assisted instruments. An example for this is given in Fig. 1.8. As one can see here, many of the extended sources identified by Becklin & Neugebauer (1975) show up as dense (in projection) groups of stars. Indeed, K-band maps of the central cluster show a tightly packed system composed of more than 10,000 individual K<19[2] stars within the central $\sim$1.5 pc. Dust and gas structures, like the mini-spiral, become observable only when observing in longer wavelength bands, such as the L-band (3.8 µm) and longer wavelength bands. Fig. 1.5 shows an example of this, where extended emission can be seen only in the red (L-band) channel.

From analysis of the distribution of stars within the cluster, it has been found that the surface density happens to peak at the position of SgrA*. As one can see in Fig. 1.9, the surface density as a function of projected radius p from SgrA* is well described by a broken power-law with a break radius of $\sim$10". After deprojection from p to the physical radius r, Genzel et al. (2003a) found a number density distribution $n(r) \propto (r/10'')^{-\alpha}$ with α=1.4 for r <10" and α=2.0 for $r \geq$10". More recently, Schödel et al. (2007) used star counts, to which a more sophisticated extinction correction was applied, for computing the surface density distribution. This analysis found basically the same results. A model of a flattened isothermal sphere fits well the outer parts ($p \geq$10") of the density distribution but reveals a cusp of stars in the innermost few arcsec.

The central cluster is composed of two main stellar populations of significantly different stellar types. Most of the stars are evolved (mean age $\sim$12 Gyr) late type giant stars showing CO absorption features in their spectra. Late type supergiants, like IRS7 or IRS19, are rarely observed. The second small ($\sim$100 members) group of stars is composed of young ($\sim$6 Myrs) OB supergiant, giant, and main sequence stars. Prominent examples of this group are the IRS16 stars whose strong He I emission was already reported by Krabbe et al. (1991). These early-type stars are more concentrated towards in the inner cusp (see Fig. 1.9).

Photometric and spectroscopic surveys (Blum, Sellgren & DePoy 1996a,b; Ott, Eckart & Genzel 1999) found several variable stars within both populations, in the late-type ensemble (e.g. IRS1W, IRS7) as well as in the early-type ensemble (e.g.

---
[1]"IRS" simply stands for "infrared source".
[2]Unless explicitly modified, all magnitudes (like H, K, L, ...) in this thesis are apparent magnitudes.

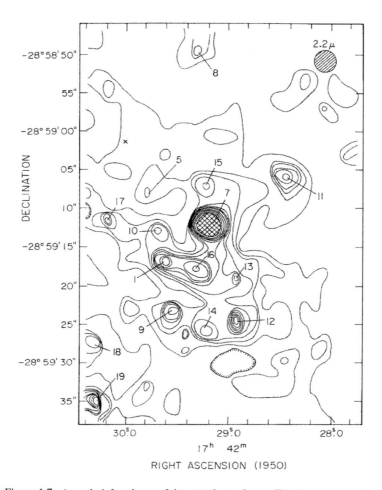

Figure 1.7: An early infrared map of the central star cluster. This image was obtained by Becklin & Neugebauer (1975) in K-band, achieving a spatial resolution of ∼2.5" (the beam size is given in the upper right corner). Individual sources are labeled with numbers; this nomenclature is still in use (cf. Fig. 1.8).

1.1. The Galactic Center

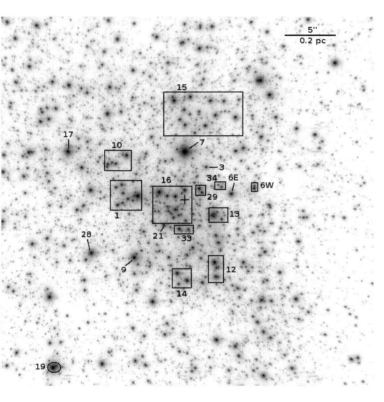

Figure 1.8: K-band image of the central star cluster; north is up, east to the left. This map is a mosaic of images obtained in 2005 at the 8.2-m VLT using adaptive optics, achieving a diffraction-limited resolution of ∼60 mas. The most important sources are labeled with their IRS numbers. Within groups of stars, individual objects can be recognized by refinements of the respective IRS labels; see e.g. IRS6E/W or Fig. 4.4. The location of SgrA* is marked by the cross in the IRS16 group.

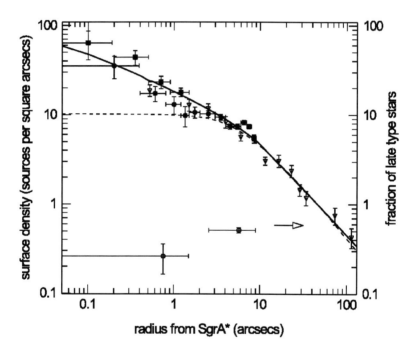

Figure 1.9: Surface density of stars (filled squares and open triangles) and fraction of late type stars (open circles) vs. projected radius from SgrA*. The surface density shows a broken power-law profile with a break radius of ∼10". A comparison with a model of a flattened isothermal sphere (dashed line) reveals a stellar cusp located in the innermost few arcseconds. This cusp shows a concentration of early type stars. *Figure: Genzel et al. (2003a).*

1.1. The Galactic Center

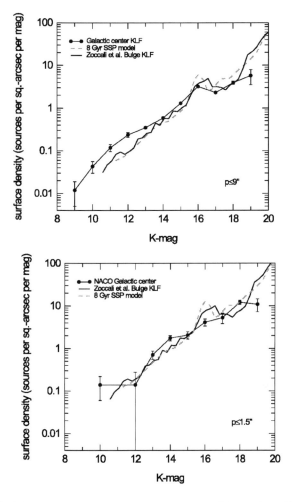

Figure 1.10: K-band luminosity functions (KLF, sources/arcsec^2/mag; points with error bars) measured for regions with projected distances p from SgrA* of $p \leq 9"$ *(top panel)* and $p \leq 1.5"$ *(bottom panel)*. For comparison, the KLF of the Galactic bulge on spatial scales of degrees (continuous line) and a single-age (8 Gyrs) stellar population model (SSP; dashed line) are shown (both from Zoccali et al. (2003)). The main difference between the two measured KLFs is a significant excess at K~16 present only for $p \leq 9"$. This excess (its absence) is due to the presence (absence) of horizontal-branch and red-clump stars in the respective parts of the GC cluster. *Figure: Genzel et al. (2003a).*

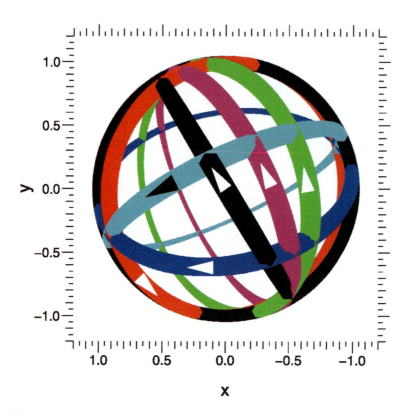

Figure 1.11: Important planar structures in the GC. North is up, east to the left, SgrA* is located at (0,0); units are arbitrary. Shown are the Galactic plane and the sky plane (*black*), the clockwise (*blue*) and counter-clockwise (*red*) early-type star disks, the northern arm (*green*) and bar (*cyan*) components of the mini-spiral, and the CND (*magenta*). The thicknesses of the rings illustrate the proximity to the observer, arrows indicate the directions of rotation. *Figure: Paumard et al. (2006)*.

IRS16SW, IRS34W). Although IRS16SW was already described and classified as an eclipsing binary by Ott et al. (1999) based on the star's K-band lightcurve, the final proof of binarity was given only recently by Martins et al. (2006) using additional radial velocity information. So far, IRS16SW is the only binary detected in the GC cluster. The case of IRS34W is discussed in detail in chapter 2.

Analyzing the cluster's K-band luminosity function (KLF; source number per area per luminosity, see Fig. 1.10) leads to the conclusion that its stellar content is in general well described as a $\sim$8 Gyr old single-age population. A systematic excess of measured star counts compared to expected values at the high-luminosity end of the KLF can be interpreted to be due to the admixture of young, hot early-type stars. Additionally, the observed KLF shows a local maximum at K$\sim$16 which is also present in both, the large scale (several degrees) Galactic bulge population and the theoretical curves, but only in the outer parts ($p > 1.5$") of the cluster. This excess (its absence) indicates the presence (the absence) of horizontal-branch and red-clump stars in the respective regions of the central cluster (Genzel et al. 2003a).

In contrast to the late-type population, the early-type stars show a very peculiar geometry in their distribution within the cluster. One group of stars, the S-stars, is concentrated in the immediate vicinity ($p < 1$") of SgrA*. Their presence so close to the central mass, which should prevent any star formation due to tidal disruption effects, is hard to understand and thus forms the "paradox of youth" (Ghez et al. 2003). Another group of early-type stars (with $p < 0.5$ pc) is arranged in two counter-rotating disks centered on SgrA*. For both groups of stars proposed explanations for their presence and distribution are divided into two classes, which are (a) in-situ-formation models and (b) infall or migration models (Paumard et al. 2006, and references therein).

Fig. 1.11 gives an overview over the relevant planar structures in the GC region. The Galactic and sky planes aside, the most important structures are the early-type star disks and the planes spanned by the main components of the mini-spiral and the CND.

1.1.3 Sagittarius A*

The attempt to examine the properties of SgrA* in the near infrared is immediately challenged by the fact that this object is a radio source and – in general – invisible in other wavelength regimes. Therefore, localizing SgrA* in infrared images requires a cross-comparison with radio data. Fortunately, the GC cluster contains a group of nine SiO maser stars which are observable in both, infared and radio bands. Correlating their observed radio and infrared positions allows the infrared position of SgrA* to be determined. This shows that it is located in the center of the S-star group (cf. Fig. 1.6), with an accuracy of $\sim$1 mas; for details, see chapters 4 and 5.

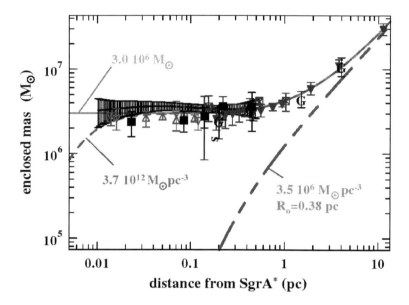

Figure 1.12: Enclosed mass vs. projected distance from SgrA*, assuming a GC distance of 8 kpc. Data points with error bars mark values obtained from stellar and gas (labeled "G") dynamics. The long-short-dashed line corresponds to the visible star cluster. The short-dashed curve represents the visible cluster plus a very concentrated cusp of compact stellar remnants. The continuous line, which reproduces the data best, corresponds to a sum of the visible star cluster and a central $3 \cdot 10^6 \, M_\odot$ point mass. *Figure: Genzel et al. (2000)*.

The possibility that the center of the Milky Way might host a supermassive black hole was suggested already by Lynden-Bell & Rees (1971) three years before the discovery of the radio source SgrA* by Balick & Brown (1974). Direct tests of the mass distribution in the GC region became possible when high-resolution infrared imaging and spectroscopy of the central cluster allowed the measurement of proper motions and radial velocities of stars in the innermost arcseconds of the galaxy (Krabbe et al. 1995; Eckart & Genzel 1997; Ghez et al. 1998; Genzel et al. 1996, 1997, 2000). These data could be converted into mass estimates using various types of statistical estimators.

All proposed mass estimators are based on the virial theorem and assume a system of stars being in dynamical equilibrium. In the most direct form, the total mass M of the stellar system can be estimated via the relation

$$G \cdot M = \frac{\langle v^2 \rangle}{\langle 1/r \rangle} \tag{1.1}$$

1.1. The Galactic Center

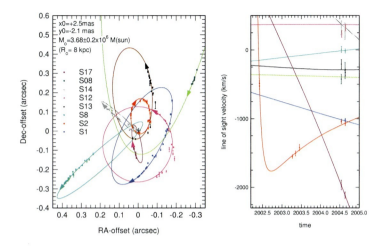

Figure 1.13: Stellar orbits around SgrA*. Points with error bars are measured time-resolved positions in the plane of sky (*left hand panel*) and line-of-sight velocities (*right hand panel*) for 8 stars. Continuous lines indicate the respective Keplerian orbital solutions. In the data set shown here, the best-fitting orbits correspond to a black hole mass of $3.7 \cdot 10^6 \, M_\odot$. *Figure: Eisenhauer et al. (2005a).*

Here G is Newton's constant, v the stellar velocity, and r the distance from the center of mass; $\langle \cdot \rangle$ denotes the average over all stars.

As this simple virial mass estimator has problematic statistical properties like bias (discussed in detail by Bahcall & Tremaine 1980), a number of more sophisticated estimators have been proposed (Bahcall & Tremaine 1980; Heisler, Tremaine & Bahcall 1985; Leonard & Merrit 1989). Nevertheless, one has to keep in mind that all these estimators make implicit assumptions on homogeneity, isotropy, average eccentricity of star orbits and other parameters. These assumptions are expressed in various geometric factors which are introduced by the treatment of projection.

The main results of statistical mass estimates for the GC cluster are summarized in Fig. 1.12. Using a GC distance R_0 of 8 kpc, the best explanation for the observed mass distribution is a superposition of a smooth mass contribution corresponding to the visible star cluster and a central ($p \leq 0.01$ pc) point mass of $\sim 3 \cdot 10^6 \, M_\odot$.

If one has proper motions and radial velocities for a sufficient number of stars, it is possible to compute R_0 as a statistical parallax. According to Salim & Gould (1999), "the distance to the center of the Galaxy (R_0) is to Galactic astronomy what the Hubble constant (H_0) is to extragalactic astronomy and cosmology". Obtaining

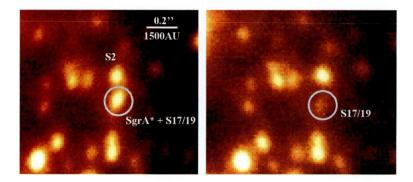

Figure 1.14: K-band flare from SgrA* as observed in May 2006. The time difference between the two images is 81 minutes. At the epoch of this observation two faint stars, S17 and S19, were located at the projected position of SgrA*. This event is described in detail in chapter 3.

an accurate value for R_0 has a high impact on understanding structure and dynamics of the Milky Way (e.g., Dehnen & Binney 1998). The statistical parallax is based on the assumption that the velocities of the observed stars are distributed isotropically. In this case, the velocity dispersions in the plane of the sky and the dispersion of the radial velocities are equal. As proper motions are measured in angular units (usually mas/yr) and radial velocities in physical units (usually km/s), R_0 is given as the scaling factor between these values. The dynamical studies mentioned above found values of $R_0 = 8 \pm 1$ kpc. This value is similar to the result $R_0 = 8.0 \pm 0.5$ kpc that had already been obtained by Reid (1993); it was found by combining estimates from a variety of independent studies.

The next development in the determination of the mass of and distance to SgrA* was the observation of Keplerian orbits (see Fig. 1.13) around the central mass (Schödel et al. 2002, 2003; Ghez et al. 2005a; Eisenhauer et al. 2005a). Using a sufficient number of time resolved positions and radial velocities for a star showing significant acceleration, it is possible to find a unique orbital solution directly delivering mass of and distance to the black hole. Technically, the star's orbit is found by a χ^2-fit of a Keplerian orbit to the data. The best test star used for determinations of mass of and distance from SgrA* is S2, which has an orbital period of 15.2 years; the first orbit fully covered by observations was completed in 2007.

The most recent values obtained so far are a mass of $M_\bullet = 3.6 \pm 0.3 \cdot 10^6 \, M_\odot$ and a GC distance of $R_0 = 7.6 \pm 0.3$ kpc (Eisenhauer et al. 2005a). Due to the geometry of the problem $M_\bullet$ strongly depends on R_0 via $M_\bullet \propto R_0^\gamma$ – with γ ranging from 0 (if the mass estimate is entirely based on radial velocities) to 3 (if the mass estimate is based

1.1. The Galactic Center

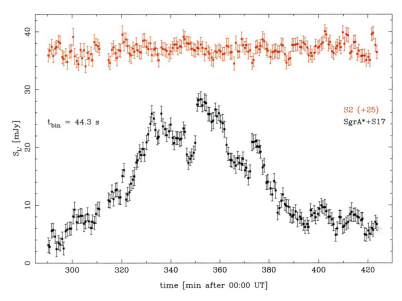

Figure 1.15: Lightcurve of a bright L-band flare obtained in April 2007. Black data points are the fluxes of SgrA* (on top of S17), red data points are the fluxes of S2 for comparison. The ∼20-min substructure in the flare is clearly visible.

on astrometric positions only).

An important point is that SgrA* is spatially resolved in the short-wavelength end of the radio regime. At $\lambda = 3.5$ mm its intrinsic size is about 1 AU or ∼15 Schwarzschild radii (Bower et al. 2004; Shen et al. 2005).

Although SgrA* is primarily a radio source, it occasionally shows outbursts ("flares") of radiation in other wavelength bands from NIR to X-rays; for a detailed overview, see chapter 3. In near infrared, flaring activity was first reported by Genzel et al. (2003b). Since then, several ten infrared events like the one presented in Fig. 1.14 have been observed by several groups (Ghez et al. 2004, 2005b; Clénet et al. 2004, 2005; Eisenhauer et al. 2005a; Eckart et al. 2006a,b; Yusef-Zadeh et al. 2006a; Krabbe et al. 2006; Hornstein et al. 2007). This large set of observations allows the derivation of some general properties of infrared flares.

Emission from SgrA* is *flaring*. In general, a flare rises from non-detection to peak fluxes up to ∼10 times the background level within ∼10...30 minutes and declines to undetectable levels again afterwards.

Emission from SgrA* shows three *characteristic timescales*: (1) A typical flare event rate of 3-4 events per day, (2) a characteristic flare duration of 1-3 hours, and (3)

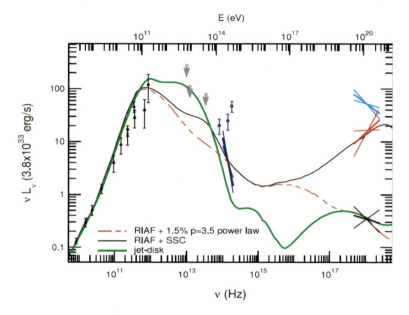

Figure 1.16: Spectral energy distribution (SED) of emission from SgrA*. Data points with errors are flux measurements from radio to infrared. Lines with error lines are flux and spectral slope measurements obtained in NIR and X-ray bands. Continuous and dashed curves are theoretical SEDs calculated from different types of emission models. *Figure: Eisenhauer et al. (2005a)*.

quasi-periodic substructure superimposed on the overall flare profiles with timescales of 15-20 minutes. An example is given in Fig. 1.15 which shows the lightcurve of an L-band flare observed in April 2007; this event was one of the most luminous L-band outbursts ever detected. If one identifies the 15-min substructure (point 3) with material orbiting SgrA* close to the innermost stable circular orbit (ISCO) within a Kerr metric, this constrains the Kerr spin parameter to $a \geq 0.6$.

Emission from SgrA* is *polarized* with polarization fractions up to ~40%. This polarization confirms the synchrotron nature of the radiation.

Combining flux measurements obtained over a wide range of wavelength bands from radio to X-ray results in an approximated spectral energy distribution (SED). This is presented in Fig. 1.16 showing measured fluxes and spectral slopes together with a set of emission model curves. In the NIR, (bolometric) flare luminosities are $\nu L_\nu \sim 10^{28}$ W ($\sim 100 L_\odot$).

The emission from SgrA* is understood to be synchrotron emission from relativistic electrons close to the ISCO. The best agreements with the data are achieved by radiatively inefficient accretion flow (RIAF) models with a thermal ($T_e \sim 10^{11}$ K) electron population. A small part of the electrons (a few %) is scattered up to even higher energies (several GeV, corresponding to $\sim 10^{14}$ K) via synchrotron self-compton (SSC) scattering processes (e.g. Genzel et al. 2003b; Yuan et al. 2004; Liu et al. 2004; Broderick & Loeb 2006; Hamaus et al. *in prep*).

1.2 Observations

The results discussed in this dissertation are based on near-infrared observations obtained at three observatories located in Chile. These are the 2.2-m MPG[3] telescope and the 3.5-m ESO[4] New Technology Telescope (NTT), both located in La Silla, and the 8.2-m Unit Telescopes (UT) which are part of the Very Large Telescope (VLT) observatory on Cerro Paranal. Additionally, a public data set obtained at the Gemini North observatory located on Mauna Kea, Hawaii, was used.

1.2.1 Obtaining diffraction-limited data

Theoretically, the angular resolution that can be achieved with a given telescope is limited only by the diffraction at the telescope aperture. In the case of a circular aperture a point source observed at infinity is mapped into an Airy profile. The spatial resolution of the optical system is defined as the angular radius of the first diffraction minimum, which is

$$\Theta = 1.22 \cdot \frac{\lambda}{D} \qquad (1.2)$$

Here λ is the observed wavelength and D is the telescope aperture. Closely connected to this parameter is the full width at half maximum (FWHM) of the central maximum of the Airy profile, which is given by

$$\Theta' = 0.98 \cdot \frac{\lambda}{D} \qquad (1.3)$$

The FWHM is commonly used to quantify telescope performances and data qualities. To give some examples: when observing in K-band ($\lambda = 2.2$ μm), diffraction-limited resolutions (FWHM) are ~ 0.20" for $D = 2.2$ m, ~ 0.13" for $D = 3.5$ m, and ~ 0.054" for $D = 8.2$ m.

---

[3]Max-Planck-Gesellschaft, Munich, Germany
[4]European Southern Observatory, Garching, Germany

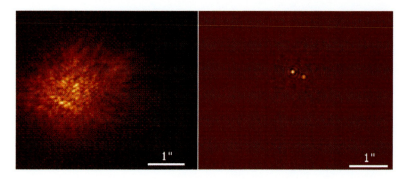

Figure 1.17: Speckle imaging observation of a binary star. *Left hand panel*: speckle pattern of the target source obtained with a few milli-second long single exposure. *Right hand panel*: Diffraction limited resolution image of the binary as reconstructed from 378 individual exposures. *Image: MPIfR*.

Unfortunately, the angular resolution of ground-based telescopes is severely limited by the influence of the atmosphere, which is highly turbulent. Due to local temperature differences, the atmosphere fragments into areas (*turbulence cells*) of different density, leading to light deflection. Planar wavefronts arriving from astronomical objects become distorted, images of these objects obtained from the ground are blurred. This effect is called *seeing*.

The typical size of a turbulence cell is described by *Fried's parameter* r_0. By replacing D by r_0 in the equations given above, one obtains the seeing-limited resolution that can be achieved by ground-based observatories. With r_0 having typical values $\sim$1 m in the near infrared, one finds a typical seeing of $\sim$0.5". As Fried's parameter depends on the wavelength like

$$r_0 \propto \lambda^{6/5} \qquad (1.4)$$

one obtains a weak wavelength dependence of the seeing following

$$\Theta_{\text{seeing}} \propto \lambda^{-1/5} \qquad (1.5)$$

The time scale for which the wavefront distortion is stationary is given by the *coherence time* τ_0. Fried's parameter and the coherence time are connected via the typical velocity of turbulence cells in the atmosphere v_0 like

$$\tau_0 = 0.38 \cdot \frac{r_0}{v_0} \qquad (1.6)$$

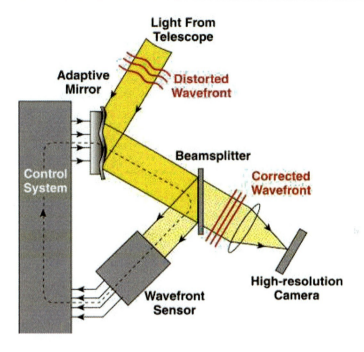

Figure 1.18: Sketch of an adaptive optics system. The infalling wavefront is analyzed by a wavefront sensor. The distortions are corrected by a deformable mirror in the light path. *Figure: IfA, University of Hawaii.*

(Roddier et al. 1982; Smith 1995; Brandl 1996; Léna, Lebrun & Mignard 1998).

In order to overcome the limitations on the angular resolution due to the seeing, it is necessary to remove the wavefront distortions imprinted on images of astronomical objects. Today, two technologies for obtaining diffraction-limited images with large ground-based telescopes are available: (1) speckle interferometry and (2) adaptive optics.

The concept of *speckle imaging* makes use of the fact that the wavefront distortion can be taken as stationary for time spans not exceeding τ_0. A short-time ($< \tau_0$) exposure of a point source is decomposed into a pattern of $N \approx D/r_0$ grains or *speckles* with a minimum grain size of $\sim \Theta$. By obtaining a sufficient number (at least several hundred) of individual short-time images it is possible to construct a diffraction-limited exposure of the target source (Labeyrie 1970). For this image reconstruction several methods combining the individual short-time exposures either in real space or in Fourier space are known. In this section I briefly discuss the simple-shift-and-add

(SSA; e.g. Christou 1991) method which was used for our speckle imaging data.

The SSA algorithm combines the recorded short-time images in real space. It requires the presence of a dominating bright point source in the field of view (FOV) which delivers a prominent reference speckle in each exposure. The individual frames are shifted such that the maxima (or centroids) of the reference speckles are located on top of each other. After this, the individual images are coadded. The resulting combined map is a diffraction-limited image of the target source. An example for this is given in Fig. 1.17; shown here is the speckle image reconstruction of the image of a binary star by combining 378 individual short-time exposures.

The concept of *adaptive optics* (AO) makes use of the fact that the theoretical, diffraction-limited image of a point source is known and can be compared to the observed, seeing limited image of the same source (Bourdet et al. 1978). This comparison allows to calculate the wavefront distortion. Diffraction limited data are thus obtained by permanently observing a sufficiently bright reference star in the target FOV and calculating the respective wavefront distortions. The distortions are compensated by a deformable mirror in the light path between telescope and camera. As the procedure of wavefront analysis and correction has to be completed within the coherence time, fast correction rates of several 100 Hz are necessary.

Fig. 1.18 gives an overview over the main components of an AO system. The infalling light is reflected onto a detector by a deformable mirror. A certain fraction of the light is fed into a wavefront sensor which analyzes the distortion. A control system computes the necessary correction and translates it into a mirror deformation. The deformable mirror compensates the wavefront distortion.

1.2.2 Instruments

The data presented and analyzed in this dissertation were obtained with five different instruments, four of these were at least partially constructed at the MPE.

The speckle imaging camera SHARP I (Fig. 1.19) was in operation from 1992 to 2002. It was designed and constructed at the MPE to obtain diffraction-limited J, H, and K-band images of the Galactic Genter (Hofmann et al. 1992). SHARP I was equipped with a 256×256 pixel NICMOS 3 detector. It was sensitive to wavelengths in the range 1.0-2.5 µm. Mounted at the 3.5-m NTT, it achieved angular resolutions of $\sim$150 mas with a detector plate scale of 50 mas/pixel. SHARP I could obtain imaging, polarimetric, and spectroscopic data. Its main optical components were 12 broad-band and narrow-band filters, a BaF_2 wire grid polarizer, and a dual prism allowing for a spectral resolution of 35 in K-band.

The integral field spectrometer 3D (Weitzel et al. 1996) was used for GC observa-

1.2. Observations

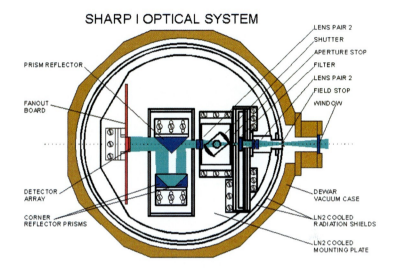

Figure 1.19: Optical system of the speckle imaging camera SHARP I. This camera obtained sets of short-time exposures with exposure times down to 0.15 sec. It was used to obtain diffraction-limited NIR images of the GC from 1992 to 2002. *Figure: T. Ott/MPE.*

tions in 1996. It was designed and built at the MPE and was one of the first integral field units ever constructed.

The basic concept of integral field spectroscopy is illustrated in Fig. 1.20. A 2-dimensional object image is sliced into thin stripes which are rearranged to a 1-dimensional pseudo-slit. This rearrangement is done optically by a system of mirrors, the "image slicer". The pseudo-slit image is dispersed using an optical grating, the resulting light distribution is imaged onto a 2-dimensional detector. The detector image is rearranged in software into a 3-dimensional data cube showing the original object image in the (x,y) plane and a spectrum for each image pixel along the z axis. Thus this method allows to obtain the full spectral information about a target object without losing the spatial information.

Like SHARP I, 3D was equipped with a 256×256 pixel NICMOS 3 detector sensitive to wavelengths in the range 1.0-2.5 μm. Mounted at the 2.2-m MPG telescope (Fig. 1.21) it obtained seeing limited data sets with a spatial plate scale of 300 mas/pixel. Depending on the actual instrument configuration, the spectral resolution was $\lambda/\Delta\lambda \approx$1000–2000. The final data cubes had the dimensions $16 \times 16 \times 256$ pixels in x, y, and z, respectively.

The NIR camera system NAOS/CONICA, NACO for short, has been in operation

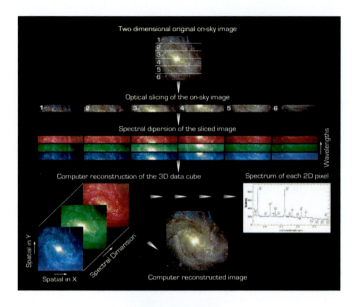

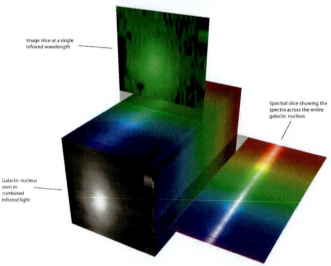

Figure 1.20: Principles of integral field spectroscopy. *Top image*: diagram showing the conversion of a 2-dimensional image into a 3-dimensional data cube. This data cube contains the full spectral information obtained from the source light while the spatial information is preserved. *Bottom image*: sketch of a 3-dimensional integral field spectroscopy data set. This data cube shows a galaxy image as an example. *Images: ESO*.

1.2. Observations

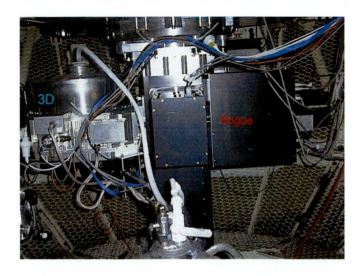

Figure 1.21: The integral field spectrometer 3D. In this image 3D (on the left) was mounted to the Cassegrain focus of the 3.9-m Anglo-Australian Telescope, together with the tip-tilt image motion corrector ROGUE (on the right). *Image: L. Tacconi-Garman/MPE.*

at the VLT (see Fig. 1.22) since 2002. NACO is a combination of the adaptive optics system NAOS (Rousset et al. 2003) and the camera CONICA (Hartung et al. 2003b). NAOS was constructed at the Observatoire de Paris; it has the unique feature of an infrared wavefront sensor allowing the use of infrared-bright but optically faint reference stars like IRS7 in the Galactic Center.

CONICA was built in collaboration by MPE and MPIA[5]. It is equipped with a 1024 × 1024-pixel ALADDIN 3 InSb detector array sensitive in the wavelength range 1-5 µm, thus covering the J to M bands. CONICA offers a variety of observing modes. Imaging observations are possible using 34 different filters and the pixel scales 13 mas/pixel, 27 mas/pixel, or 54 mas/pixel. Polarimetric imaging is done using either wire grid or Wollaston prism polarizers. Long-slit spectroscopy modes obtain spectral resolutions of 60–250 (prism) and 400–1500 (grisms). Additionally, coronographic, Fabry-Perot spectroscopic, and simultaneous differential imaging modes are available. The basic layout of NACO is presented in Fig. 1.23.

The instrument SINFONI is a combination of the AO system MACAO (Bonnet et al. 2003, 2004) and the integral field spectrograph SPIFFI (Eisenhauer et al. 2003b,c). MACAO was designed and constructed by ESO; it uses a wavefront sensor for visible

---

[5]Max-Planck-Institut für Astronomie, Heidelberg, Germany

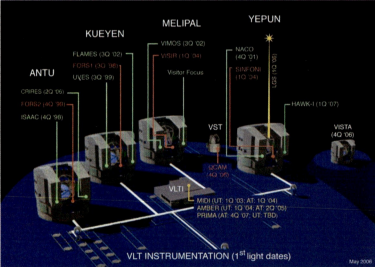

Figure 1.22: Overview images of the Very Large Telescope located on Cerro Paranal, Chile. *Top panel*: Aircraft image of the observatory obtained in 2000. *Bottom panel*: Plan of the telescopes with their respective instruments. Telescope "Yepun" hosts NACO, SINFONI, and the laser guide star system (LGSF). *Images: ESO*.

1.3. Data processing and analysis

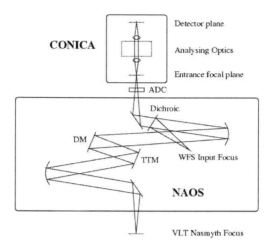

Figure 1.23: Optical design of the camera system NAOS/CONICA. NAOS is the AO system; main components are tip-tilt mirror (TTM), deformable mirror (DM), dichroic, and wavefront sensor (WFS). CONICA is the NIR camera, including the analysing optics and the detector. This instrument is mounted at the Nasmyth focus of the VLT-UT 4 Yepun. *Figure: Hartung et al. 2003a.*

light.

SPIFFI (Fig. 1.24) was built in a joint effort of MPE, ESO, and NOVA[6]. It is equipped with a 2048 × 2048-pixel HAWAI 2RG array detector, covering the wavelength range 1.0–2.5 µm (J, H, K, H+K bands). Spectral resolutions are in the range 1500–6000. SPIFFI offers three spatial image plate scales, which are 0.25", 0.1", and 0.025". The instrument has been in operation at the VLT since 2004.

Since June 2007, NAOS/CONICA and SINFONI have some of the time been operated using a sodium layer laser guide star emitted from the VLT Laser Guide Star Facility (LGSF; e.g. Rabien et al. 2003); see also Fig. 1.1.

1.3 Data processing and analysis

The output – the raw data – of astronomical instruments like those described in the previous section is the basis of any observational result. However, the gain of reliable quantitative results requires a number of calibration and analysis steps to be applied to the raw data. The main steps relevant for the data used in this dissertation I will discuss

---
[6]Nederlandse Onderzoekschool voor Astronomie, Amsterdam – Groningen – Leiden – Utrecht – Nijmegen, The Netherlands

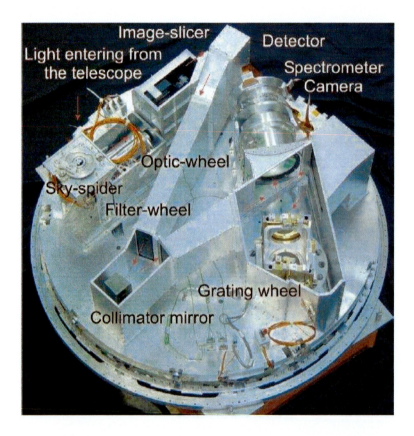

Figure 1.24: View into the integral field spectrometer SPIFFI. The main parts of the instruments are labeled, the system has a diameter of about 1 meter. *Image: R. Abuter/MPE.*

1.3. Data processing and analysis 31

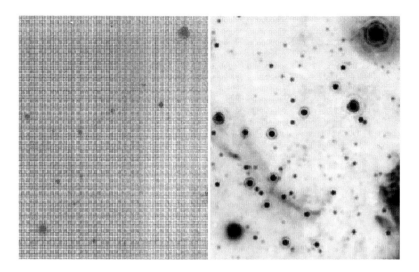

Figure 1.25: An illustration of the necessity of data reduction. *Left hand panel*: raw L-band image of the IRS16 region obtained in March 2003. Only the brightest sources are detectable by eye. *Right hand panel*: the same image after background subtraction, flat-field, and bad-pixel correction. Obviously, the data reduction process was necessary to allow for any reliable quantitative data analysis.

in this section.

1.3.1 Data reduction

Detector images show a number of effects which reduce the quality of the data set and have to be corrected.

The first important effect is the presence of *bad pixels*, i.e. detector elements which either do not respond to photon flux (dead pixels) or show a signal excess (hot pixels). Bad pixels can be identified in dark or homogeneously illuminated detector frames by comparison with neighbouring pixels. When identified, their values in a given image are replaced by fluxes obtained from surrounding pixels by interpolation. For the instruments described in the previous section, the fraction of bad pixels is (or was) in the range $\sim$0.5–5%.

The light received by an instrument is not only the radiation from the target object but a superposition of emissions from a number of additional sources. The total flux recorded by a detector, I_{total}, is given by

$$I_{\text{total}} = (I_O + I_{\text{sky}} + I_{\text{th}}) \cdot f + I_{\text{bias}} \qquad (1.7)$$

with I_O being the target object light, I_{sky} the sky background, and I_{th} the thermal background from the telescope and the instrument itself. f is the *flat field*, i.e. the relative detector sensitivity function; I_{bias} is the detector bias composed of dark current and readout noise. The science target light can be extracted from the total recorded flux using the following general scheme.

1. Obtain a *sky frame* using the same optical setup (integration times, plate scales, number of detector readouts, polarization angles, etc.) as for the science frames. This sky frame is described by

$$I_{\text{skyframe}} = (I_{\text{sky}} + I_{\text{th}}) \cdot f + I_{\text{bias}} \qquad (1.8)$$

   Subtracting the sky frame from the science images leaves

$$I_{\text{res}} = I_{\text{total}} - I_{\text{skyframe}} = I_O \cdot f \qquad (1.9)$$

   Thus the residual image I_{res} is the true object image weighted with the flat field.

2. Obtain a set of flat field calibration images composed of on- and off-frames. *On-frames* are images homogeneously illuminated with intensity I_L and are given by

$$I_{\text{on}} = f \cdot I_L + I_{\text{bias}} \qquad (1.10)$$

   Off-frames are non-illuminated images described by

$$I_{\text{off}} = I_{\text{bias}} \qquad (1.11)$$

   Subtracting the off-frames from the on-frames and using the normalization $I_L \equiv 1$ one obtains the flat field f via

$$I_{\text{on}} - I_{\text{off}} = f \cdot I_L \quad \longrightarrow \quad f = \frac{I_{\text{on}} - I_{\text{off}}}{\langle I_{\text{on}} - I_{\text{off}} \rangle} \qquad (1.12)$$

   with $\langle \cdot \rangle$ denoting the median over all pixels.

3. Extract the true object image via

1.3. Data processing and analysis

$$I_\mathrm{O} = \frac{I_\mathrm{res}}{f} = \frac{I_\mathrm{total} - I_\mathrm{skyframe}}{f} \qquad (1.13)$$

By applying the bad-pixel correction to the resulting image, one obtains the final data set.

The data processing steps described so far are summarized by the terms *image calibration* or *detector calibration* (for a detailed review, see e.g. Berry & Burnell 2000). Their execution is necessary for all types of data recorded on detectors; an example for this is given in Fig. 1.25. In case of imaging data, the data reduction is complete at this point.

In case of data sets containing *spectroscopic* information, additional steps are required. In the following, I give a brief description of the reduction of integral field spectroscopy data.

After applying bad-pixel correction and flat-fielding as described above, the wavelength axis has to be calibrated in order to reproduce the absolute positions of spectral features correctly. Also the spectral properties of the atmosphere have to be taken into account. The final data cubes are created the following way.

1. Find the function describing the instrument distortion, e.g. via the "north-south-test". In this test, continuous light fibre spectra are taken repeatedly with the fibres being moved perpendicular to the slitlets. The distortion is computed by comparing theoretical and observed fibre image positions. The transformation from raw to undistorted coordinates is usually expressed as a 2-dimensional polynomial.

2. Obtain a spectral reference image by observing the light from an emission line gas lamp; an example is shown in Fig. 1.26. The true line positions (in wavelength units) are known from atomic data catalogues. The line image positions on the detector (in pixels) are obtained by polynomial or Gaussian profile fits. By applying a polynomial fit to the observed positions, dispersion coefficients are computed. Finally, the detector rows are remapped along the spectral axis such that their effective dispersion coefficients are equal.

3. Rearrange the detector pixels in software such that the 2-dimensional detector image is mapped into a 3-dimensional data cube. The final cube stores the spatial object image in the x,y plane. Along the z axis the spectral information is recorded.

4. Obtain a sky cube using the same optical setup (filters, integration times, plate

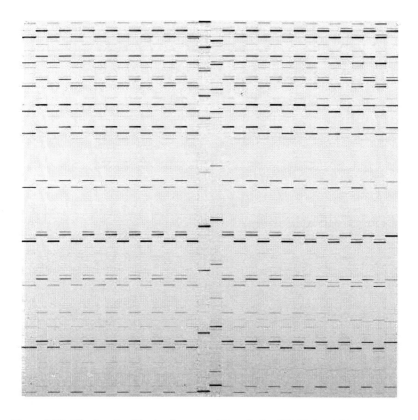

Figure 1.26: Wavelength calibration image used for reducing SPIFFI data. This image was obtained by observing the light from an emission line gas lamp. The spatial information (pseudoslit) is stored in the horizontal axis. The spectral information is stored in the vertical axis. Each short horizontal stripe corresponds to the image of a slitlet for the respective wavelength. After proper calibration, all slitlet images belonging to a given wavelength are aligned in a horizontal straight line spanning the full detector width.

scales, spectral resolutions, ...) as for the object cubes. Subtracting the sky from
the object cubes removes the atmospheric *emission*.

5. Take a standard star cube to record the atmospheric *absorption*. The atmosphere's absorption spectrum can be obtained by dividing the observed star spectrum by a theoretical one. Dividing the observed object cubes by the absorption spectrum produces the true object cubes.

After executing the steps given above, the resulting data cubes are ready for analysis (for the case of SPIFFI, see e.g. Abuter et al. 2006, Modigliani et al. 2007).

1.3.2 Deconvolution

Although the imaging data used in this dissertation have been corrected for atmospheric seeing by speckle imaging or adaptive optics techniques, the final object images are still far away from the theoretical limits. This is caused mainly by the fact that the speckle or AO correction cannot be perfect. It restores the diffraction-limited object images only partially, leading to Strehl ratios in the range $\sim$20–50% depending on the actual atmospheric conditions (especially Θ_{seeing}, τ_0).

The observed image of a point source can usually be described as the superposition of a diffraction-limited core and an extended (to about Θ_{seeing}) seeing halo. More generally, the observed source image $I(\mathbf{x})$ is given by

$$I(\mathbf{x}) = O(\mathbf{x}) * P(\mathbf{x}) + N(\mathbf{x}) \tag{1.14}$$

with $O(\mathbf{x})$ being the true object image, $P(\mathbf{x})$ being the point spread function (PSF) describing the image blur, $N(\mathbf{x})$ being additive noise, "$*$" denoting convolution, and $\mathbf{x}$ being the position in the image plane. In the following, upper-case characters denote parameters in image space, lower-case letters the respective Fourier transforms in frequency space with coordinates $\mathbf{u}$. The problem of restoring $O(\mathbf{x})$ using a given PSF is known as the *inverse convolution problem* or *deconvolution problem*.

The following way to extract a PSF from a given image has been used extensively throughout this dissertation. From the image of a star cluster (in our case, of course, the GC cluster) a set of about ten bright, isolated stars is selected. Each star is cut out of the image using a circular aperture and normalized to unit peak flux. After this, all star images are median combined. The resulting image represents a typical stellar PSF.

According to the convolution theorem, the Fourier transform of the observed object image is given by

$$i(\mathbf{u}) = o(\mathbf{u}) \cdot p(\mathbf{u}) + n(\mathbf{u}) \tag{1.15}$$

Thus, when ignoring noise, one might be tempted to restore the true object image by a direct Fourier division $o(\mathbf{u}) = i(\mathbf{u})/p(\mathbf{u})$. Unfortunately, this ansatz is unusable in general as $p(\mathbf{u})$ needs to be non-zero for all $\mathbf{u}$, meaning that the PSF must provide information for all frequencies. Ultimately, the presence of noise obviates a unique solution.

As a *unique* solution for $O(\mathbf{x})$ usually cannot be found, several deconvolution algorithms have been developed which reconstruct the *most probable* solution. In the following, I give a description of those algorithms which were used at least occasionally during this dissertation.

The *Wiener algorithm* (Wiener 1950; Bates, Fright & Bates 1984; Ott et al. 1999) uses a non-iterative multiplicative scheme. It reconstructs the true object image using the filter function

$$w(\mathbf{u}) = \frac{p^*(\mathbf{u})}{\mid p(\mathbf{u}) \mid^2 + \mid \tilde{\delta}(\mathbf{u}) \mid^2} \qquad (1.16)$$

where $p^*(\mathbf{u})$ is the complex conjugate of $p(\mathbf{u})$ and $\tilde{\delta}(\mathbf{u})$ is the the Fourier transform of Dirac's δ function.

The object image is restored via $o(\mathbf{u}) = i(\mathbf{u}) \cdot w(\mathbf{u})$ and Fourier transformation of the result. Obviously, this scheme would correspond to a direct Fourier division in case of a vanishing $\tilde{\delta}(\mathbf{u})$.

The *Lucy-Richardson algorithm* (Richardson 1972; Lucy 1974) uses an iterative scheme. Each iteration is given as

$$O_{i+1}(\mathbf{x}) = O_i(\mathbf{x}) \cdot \left[P(-\mathbf{x}) * \frac{I(\mathbf{x})}{P(\mathbf{x}) * O_i(\mathbf{x})} \right] \qquad (1.17)$$

with $O_i(\mathbf{x})$ being the reconstructed image generated after iteration i; $O_0(\mathbf{x})$ corresponds to the original input image. The quantity $P(-\mathbf{x})$ denotes the spatially rearranged PSF. This rearrangement should be selected properly in order to catch symmetric deviations from a diffraction limited point source profile, especially symmetric elongations along one axis. For example, one might flip the image with respect to one or both axes, or rotate the image about the image center.

Depending on the actual image quality, a good image reconstruction can require several 1,00 to several 10,000 iterations. In cases where the FWHM of reconstructed point sources falls below the diffraction limit, beam restoration by convolution with a diffraction-limited beam profile should be applied.

The *CLEAN algorithm* (Högbom 1974) follows an iterative additive scheme and operates in image space. It does not use convolutions but PSF subtraction using the following steps:

1.3. Data processing and analysis

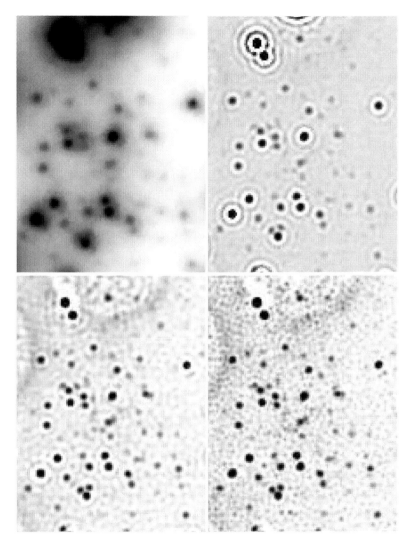

Figure 1.27: An illustration of different deconvolution techniques. *Top left panel*: The original image, an H-band map of the S-star region obtained in April 2007. The same image is shown after applying Wiener filtering (*top right panel*), Lucy-Richardson (LR) deconvolution (*bottom left panel*), or Högbom-CLEAN deconvolution (*bottom right panel*). In the cases of LR and CLEAN deconvolution, the algorithms were pushed slightly beyond the threshold of noise enhancement in order to restore even the faintest sources.

1. Find the pixel (x_{max},y_{max}) with the highest flux value f_{max} in the original image (the "dirty map").

2. Center the template PSF (the "dirty beam") on (x_{max},y_{max}), normalize its peak value to $g \cdot f_{max}$, and subtract this normalized PSF from the dirty map. Here g is the loop-gain factor used to control speed and quality of the process; it has to be smaller than or equal unity.

3. Map the subtracted flux as a δ-function into the respective position of a "clean map".

4. Repeat steps 1–3 until the residuals in the dirty map fall below a given significance threshold, i.e. convergence is reached.

5. As the clean map is composed of δ-peaks, usually beam restoration is necessary. After this, add the residual dirty map to the clean map to preserve fluxes.

This algorithm has several features which need to be taken into account. First, it centers the PSF on the brightest pixel of the residual dirty map. Thus, the number of iterations operating on the same source and thus the quality of reconstruction depends on the source brightness with respect to other sources in the field of view. Second, it decomposes the dirty map into a set of δ-functions. In cases where spatially extended or resolved sources are present in the dirty map aside with point sources, this can lead to a poor reconstruction. This "multiresolution problem" is discussed in detail by Wakker & Schwarz (1988).

In cases where accurate source reconstruction and flux conservation are not crucial, e.g. for creating finding charts or colour maps, simple image sharpening is usually sufficient. Image sharpening algorithms do not require a template PSF and use integral image operations, making them fast and easy to handle. They treat the input data as if the image was affected by a symmetric blur only. An example for an iterative image sharpening algorithm is given by the follwing relation.

$$O_{i+1}(\mathbf{x}) = O_i(\mathbf{x}) + [O_i(\mathbf{x}) - O_i(\mathbf{x}) * G_a(\mathbf{x})] \qquad (1.18)$$

Here $O_i(\mathbf{x})$ is the sharpened image generated after iteration i; $O_0(\mathbf{x})$ corresponds to the original input image. $G_a(\mathbf{x})$ denotes a radially symmetric Gaussian with FWHM a.

One should note that this scheme introduces non-physical negativities into the reconstructed image. This can be avoided by clipping the negative part of the expression [·]; the price one has to pay for this is the loss of flux conservation.

1.3. Data processing and analysis

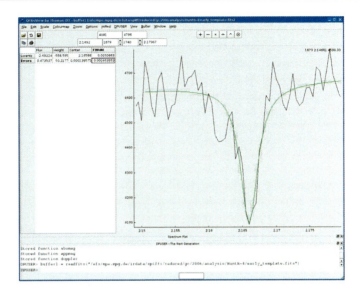

Figure 1.28: Analyzing a spectral line by fitting an analytical profile to it. In this example, flux and position of the Hydrogen Brγ line are found by fitting it with a Lorentzian function. This image is a screenshot of the MPE *QFitsView* spectrum analysis tool developed by Thomas Ott.

As convolutions are well-defined in arbitrary numbers of dimensions, all the procedures discussed in this subsection can be applied as well to 1-dimensional (spectra) or to 3-dimensional (spectroimaging data cubes) data sets.

1.3.3 Fluxes and positions

The results presented in this dissertation crucially depend on astrometric, photometric, and spectroscopic information. In order to obtain the respective data from images and spectra, a number of methods has been commonly used. All in all, the several approaches can be divided into three main groups: Counting methods, fitting methods, and correlation methods.

Counting methods are those which obtain positions and fluxes of target sources by applying counting rules to the data. In case of photometric information, this is done via *aperture photometry*. Here the flux located within a given aperture is summed up. An aperture is defined as an interval of (image or spectrum) pixels centered on a target; this target can for example be a star or a spectral line. In order to take into account possible background light (e.g. sky, nebular emission), this background is usually measured by additional apertures and subtracted from the flux found in the primary aperture. In case of images, a widely used procedure is to place an annulus aperture, which serves as background light aperture, around the target aperture. All values need to be normalized properly to take into account non-equal aperture sizes, e.g. by conversion into counts per pixel. By subtracting (normalized) background from target flux, the intrinsic target flux is obtained. Mathematically, this scheme is described by the relation

$$I = \frac{1}{N}\sum_{i=1}^{N} T_i - \frac{1}{M}\sum_{j=1}^{M} B_j \qquad (1.19)$$

I is the intrinsic target source flux per pixel, N, M are the numbers of target and background aperture pixels, respectively. T_i, B_j are target and background flux located in the i-th (j-th) aperture pixel, respectively.

Positions can be measured via *centroiding*. This scheme computes the center of light within a given aperture as the first moment of the light distribution:

$$x_{\text{centroid}} = \frac{\sum_i x_i \cdot I_i}{\sum_i I_i} \qquad (1.20)$$

Here x is an arbitrary coordinate, x_i and I_i are position and (background subtracted) flux assigned to the i-th pixel. The summation is performed over all aperture pixels (e.g. Berry & Burnell 2000).

Fitting methods obtain information by fitting the profile of an analytic function to a target object (e.g. a star in an image, a spectral line). This method is applicable if a reasonable analytic approximation of the object shape is at hand. Commonly used functions are Gaussians for stellar PSFs and Gaussians or Lorentzians for spectral lines. Typical fit parameters are position, background offset, peak height, and width. Technically, these parameters are obtained by minimizing a cost function of the well-known type

$$\chi^2 = \sum_i \frac{(x_i - m_i)^2}{\sigma_i^2} \qquad (1.21)$$

Here x_i, m_i are the measured and the model value of data point i; σ_i is the respective statistical error. A fit of this type delivers source position and source flux in one step; see Fig. 1.28 for an example (e.g. Stone 1989, and references therein).

Correlation methods rely on a quantitative comparison between the observed target and an independently obtained template. This template is usually either a theoretical model or another data set. This method is based on the assumption that observed data and template are identical, but shifted with respect to each other. Here the term "shifted" refers to a properly chosen general coordinate; frequent choices are positions

1.3. Data processing and analysis

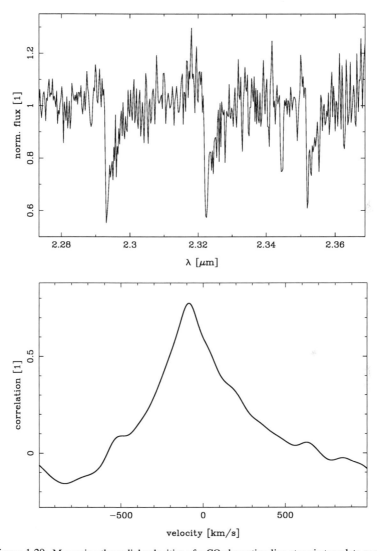

Figure 1.29: Measuring the radial velocities of a CO absorption line star via template correlation. The correlation was calculated from the three first CO band heads in the wavelength range 2.29-2.36μm (*top panel*). A theoretical model spectrum served as template. The peak of the correlation curve (*bottom panel*) corresponds to the best-fitting velocity. In this case, the star moves with −87 km/s relative to the observer.

x,y in images or wavelengths λ in spectra. For a given test value x_t of a coordinate x the correlation coefficient of source a and template b is defined as

$$c(x_t) = \frac{\sum_x [a(x) - \bar{a}] \cdot [b(x - x_t) - \bar{b}]}{\sqrt{\sum_x [a(x) - \bar{a}]^2 \cdot \sum_x [b(x - x_t) - \bar{b}]^2}} \quad (1.22)$$

(for a general definition, see e.g. Bronstein et al. 1999). The summations are performed over all coordinate entries x, e.g. all wavelength bins of a spectrum. $\bar{a}, \bar{b}$ are the means of a, b, respectively. By definition, $c(x)$ is limited to the range $[-1; +1]$; a value of 0 means absence of any correlation, whereas $+1(-1)$ corresponds to perfect (anti)correlation. Sub-bin accurate values are obtained by proper interpolation of the template. By scanning a sufficient number of test values, a correlation curve is build up. The location of the curve's peak corresponds to the best-fitting shift.

A frequent application are radial velocity measurements. Here the observed spectrum of a star is compared to a respective template (see Fig. 1.29 for an example). Another common application is the determination of accurate stellar positions in an image via star-vs.-PSF-template correlation (e.g. Diolaiti et al. 2000).

Stellar fluxes and positions obtained with the methods outlined above usually need to be converted into proper absolute physical units, e.g. photometric magnitudes (fluxes) or astrometric coordinates (positions). A common way – and the method used throughout this thesis – is the simultaneous measurement of fluxes (positions) of a target source and several reference objects with known photometric magnitudes (astrometric positions). Comparing instrumental vs. physical fluxes (positions) of the *reference* objects allows to derive scaling factors (fluxes) or coordinate transformation parameters (positions). Those parameters are then used to convert the instrumental values obtained for the *target* source into physical units.

1.3.4 Time series

Although of moderate importance for this dissertation (mainly section 3.4; but see also Genzel et al. 2003b, Martins et al. 2006, Gillessen et al. 2006, Rank 2007), I will briefly discuss the concept of time series analysis. This term summarizes several schemes used for detecting (quasi)periodic signals in time-resolved data and/or measuring their period. In the following, I distinguish two types of analysis methods: periodogram methods and phase diagram methods.

A *periodogram* of a time series $\{X\}_j$ is based on its discrete Fourier transform

$$f_X(\omega) = \sum_j X_j \cos(\omega t_j) + i \cdot \sum_j X_j \sin(\omega t_j) \quad (1.23)$$

1.3. Data processing and analysis

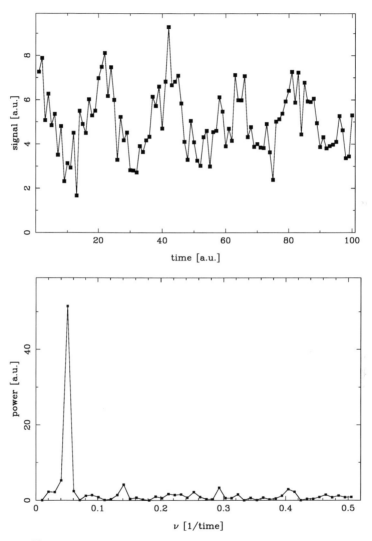

Figure 1.30: Time series analysis with a Scargle periodogram. *Top panel*: Simulated time-resolved data set. This signal contains a cosine-wave with period 21 and amplitude 1.5. Noise is Gaussian with $\sigma = 1$; all units are arbitrary. The periodicity is hardly recognizable by eye. *Bottom panel*: Scargle periodogram of the data. In spite of the poor signal-to-noise ratio, the curve shows a strong peak at the frequency of the periodic signal.

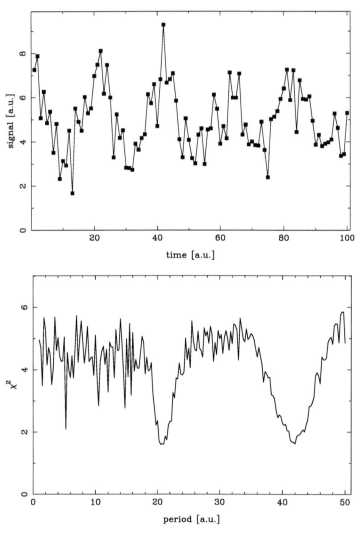

Figure 1.31: Period backfolding analysis of a time series. *Top panel*: The same simulated signal as used in Fig. 1.30. *Bottom panel*: Phase diagram of the data set. The minima indicating the signal period of 21 units and its first harmonic can be recognized easily.

1.3. Data processing and analysis

Here X_j, t_j are value and time of the j-th data point, ω is the frequency, and $i \equiv \sqrt{-1}$.

A periodogram is defined (Schuster 1898) as the squared modulus of $f_X(\omega)$,

$$A_X(\omega) \equiv | f_X(\omega) |^2 = \left(\sum_j X_j \cos(\omega t_j) \right)^2 + \left(\sum_j X_j \sin(\omega t_j) \right)^2 \quad (1.24)$$

A data set is examined by computing its periodogram for all Fourier frequencies $\omega_k = 2\pi k/T$ ($k = 1, 2, ..., N/2$, N being the number of data points, T being the length of the data set in time). If any periodicity is present in the data, the resulting ω-$A_X(\omega)$ curve shows a peak at the respective frequency; Fig. 1.30 shows an example. As $A_X(\omega)$ is very sensitive to noise, Scargle (1982) introduced the modified periodogram

$$P_X(\omega) = \frac{1}{2} \cdot \left[\frac{(\sum_j X_j \cos \omega t_j)^2}{\sum_j \cos^2 \omega t_j} + \frac{(\sum_j X_j \sin \omega t_j)^2}{\sum_j \sin^2 \omega t_j} \right] \quad (1.25)$$

This Scargle periodogram is commonly used for astronomical time series analysis.

Assigning a significance to a Scargle periodogram peak is straight forward in case the data are affected by white (i.e. Gaussian) noise with amplitude σ. For false alarm probabilites $s \ll 1$, the corresponding detection threshold is given by $z = \sigma^2 \cdot \ln(n/s)$; here n is the number of frequencies tested (Scargle 1982). In case of non-white noise – which might be an intrinsic property of the source – quantifiying significance levels requires demanding analytical (e.g. Vaughan 2005) or numerical (e.g. Timmer & König 1995; Uttley, McHardy & Papadakis 2002) approaches.

Phase diagram methods analyze a given time-series via period backfolding. For a given reference period T each data point located at time t_j is transferred to $t'_j \in [0;T]$ via $t_j \to t'_j = t_j \mod T$. For the rearranged set of N data points a χ^2 type nearest-neighbour smoothness function

$$\chi^2_{NN} = \frac{1}{N-1} \sum_{j=2}^{N} \frac{(X_j - X_{j-1})^2}{\sigma_j^2} \quad (1.26)$$

is calculated; σ_j is the error of the j-th data point X_j, $j = 1, 2, ..., N$. The smaller χ^2_{NN} is, the better is the agreement between tested and real period. By scanning a sufficient range of periods, a phase diagram is constructed (see Fig. 1.31 for an example). Minima in the χ^2_{NN} curve (if present) correspond to the locations of the signal period and its integer multiples (Lafler & Kinman 1965; Ott et al. 1999).

In contrast to periodograms, period backfolding is applicable only to data containing not more than one periodicity in an otherwise flat (except for random noise) time series. On the other hand, phase diagrams are not intrinsically limited in resolution to a set of base frequencies. For these reasons, the two schemes complement each other. Periodograms are mainly used to test if a (quasi)periodicity is present at all or if several characteristic timescales are present in the same data set (see also section 3.4). Period backfolding is usually used for precise measurements of already known (or at least suspected) periodic signals (e.g. Ott et al. 1999; Martins et al. 2006).

Chapter 2

The LBV candidate GCIRS34W

Original publication: S. Trippe, F. Martins, T. Ott, T. Paumard, R. Abuter, F. Eisenhauer, S. Gillessen, R. Genzel, A. Eckart & R. Schödel 2006, *GCIRS34W: An irregular variable in the Galactic Centre*, A&A, 448, 305

Abstract: We report the results of time-resolved photometric and spectroscopic near-infrared observations of the Ofpe/WN9 star and LBV candidate GCIRS34W in the Galactic Centre star cluster. Diffraction limited resolution photometric observations obtained in H and K bands show a strong, non-periodic variability on time scales from months to years in both bands accompanied by variations of the stellar colour. Three K band spectra obtained in 1996, 2003 and 2004 with integral field spectrometers are identical within their accuracies and exclude significant spectroscopic variability. The most probable explanation of the stellar photometric variability is obscuration by circumstellar material ejected by the star. The approximated position of GCIRS34W in a HR diagram is located between O supergiants and LBVs, suggesting that this star is a transitional object between these two phases of stellar evolution.

2.1 Introduction

The class of Ofpe/WN9 stars has received attention in the last decade. Initially defined as stars with spectroscopic characteristics of both the most extreme O supergiants and late WN stars (Walborn 1982), these objects may be in an intermediate evolutionary state between these two types of stars (Bohannan & Walborn 1989; Nota et al. 1996; Pasquali et al. 1997). Since the observation by Stahl (1983) of the Ofpe/WN9 star R127 turning to a Luminous Blue Variable (LBV), a link between these two classes of objects is suspected. Stahl (1986) also showed that the LBV AG Car displays a typical Ofpe/WN9 star in its minimum phase and Crowther et al. (1995) associated a few

Ofpe/WN9 stars with dormant LBVs from comparison of their chemical composition and physical parameters.

LBVs are well known for their strong, episodic variability (Conti 1984). These stars are usually very luminous ($L \simeq 10^6 L_\odot$). They can experience "great eruptions" with V increasing by more than 2 mag (e.g. η Car, P Cyg) or they can stay in a relatively quiescent phase with low V band variation ($\simeq 0.5$ mag) and suddenly erupt with brightening of 1-2 mag (see Humphreys & Davidson 1994 for a review). All mechanisms developed to explain the behaviour of LBVs (Glatzel & Kiriakidis 1993; Dorfi & Gautschy 2000; Humphreys & Davidson 1984; Lamers & Fitzpatrick 1988; Nugis & Lamers 2002) are associated with strong mass outflows (with $\dot{M}$ as high as $0.1 M_\odot$/yr in case of η Car, Massey 2003) which is confirmed by the detection of nebulae around LBVs (Voors et al. 2000). Hence spectroscopic variability due to both changes in T_{eff} and $\dot{M}$ is also common in LBVs.

In this article we present the results of photometric and spectroscopic long-term observations of the star GCIRS34W (hereafter referred to as IRS34W) which is part of the star cluster surrounding the supermassive black hole in the centre of our Milky Way. This star cluster, located in a distance of 7.6 kpc (Eisenhauer et al. 2005a), is a system of dynamically relaxed old stars with an admixture of young, hot, non-relaxed stars; some of the young, hot stars show strong He I emission (Krabbe et al. 1991). Most of the young and massive stars appear to reside in two thin rotating disks centered on the Galactic Centre supermassive black hole Sgr A* (Genzel et al. 2003a, Paumard et al. 2006). Among these objects, around 30 have been identified as Wolf-Rayet stars and LBV candidates (Paumard et al. 2004). Another group of early- and late-type B main sequence stars can be found in the immediate vicinity (few tens of light days) of Sgr A* (Genzel et al. 2003a, Eisenhauer et al. 2005a), leading to the so-called "paradox of youth", i.e. the question, how these massive and young stars managed to reside in an area so close to the black hole (Ghez et al. 2003).

IRS34W is one of the young He I stars and was already identified as a Ofpe/WN9 star in earlier examinations (Krabbe et al. 1995, Najarro et al. 1997); additionally, IRS34W has been classified as a LBV candidate by Paumard et al. (2001). Therefore this star is of great interest for probing the understanding of star formation and evolution in the Galactic Centre.

2.2 Observations

2.2.1 Photometry

Photometric observations of IRS34W have been obtained regularly since 1992.

From 1992 to 2002 we used the camera SHARP I (Hofmann et al. 1992) at the 3.5-m-NTT of ESO in La Silla, Chile, in speckle imaging mode to obtain diffraction limited resolution K band images. All images were sky-subtracted, bad-pixel- and flat-field-corrected and deconvolved with a Lucy-Richardson-algorithm (Lucy 1974, Richardson 1972), the spatial resolution was 130 milli-arcseconds (mas). On the processed images we applied aperture photometry, using a set of four stars located in the observed field as calibrators. Additionally in 1992, 1997 and 1998 we obtained H band images with a spatial resolution of 100 mas. These images were processed (except deconvolution) and analyzed like the K images.

Since 2002 we used the detector system NAOS/CONICA (NACO for short) consisting of the adaptive optics system NAOS (Rousset et al. 2003) and the NIR camera CONICA (Hartung et al. 2003b) at the 8.2-m-UT4 (Yepun) of the ESO-VLT on Cerro Paranal, Chile. We obtained AO-corrected diffraction limited resolution images in H (spatial resolution 40 mas) and K bands (spatial resolution 55 mas). After sky subtraction, bad-pixel- and flat-field-correction all images were deconvolved with a Wiener filtering algorithm (Ott et al. 1999). On the deconvolved images we applied aperture photometry, here using an ensemble of 12 stars in the field of view as reference sources.

Additional to our observations we included images from the Galactic Center Demonstration Science Data Set obtained in 2000 with the 8-m-telescope Gemini North on Mauna Kea, Hawaii, using the AO system Hokupa'a in combination with the NIR camera Quirc. These images were processed by the Gemini team and released to be used freely. We analyzed one H and one K band image obtained in July 2000 by applying aperture photometry calibrated with a set of three stars.

2.2.2 Spectroscopy

For a time-resolved spectroscopic analysis of IRS34W we compared three spectra obtained in 1996, 2003 and 2004.

The first spectrum was obtained in March 1996 using the integral field spectrometer 3D (Weitzel et al. 1996) at the 2.2-m-MPG-telescope in La Silla, Chile. The data output is structured as data cubes with two spatial axes, each 16 pixels long, and one spectral axis of 256 pixels length. So 3D provides seeing limited 16×16 pixel images with a spectrum for each image pixel. The March 1996 observations covered the wavelength interval from 1.97 to 2.21μm (the short wavelength half of K band) with a spectral resolution of $R = 1500$. The seeing was around 0.6", the spatial pixelscale was 300 mas/pixel. After sky subtraction, bad-pixel- and flat-field-correction the spectrum was calibrated spectrally with a neon lamp. Atmospheric features were removed by dividing by the normalized spectrum of a calibration star, only in the range of low atmospheric transmission and high noise below 2.02μm the correction was incomplete.

The second spectrum was obtained in April 2003 with the integral field spectrometer SPIFFI (Eisenhauer et al. 2003b,c) at the VLT-UT2 (Kueyen). SPIFFI produced seeing limited data cubes with 32 pixels in each spatial axis and 1024 pixels in the spectral axis. The data were corrected and calibrated (sky, bad pixels, flat field, atmosphere, wavelength calibration) like the 3D spectrum. Seeing was around 0.3", the spatial pixelscale was 100 mas/pixel. The final spectrum covered the complete K band from 1.95 to 2.45μm with a spectral resolution of $R = 3600$.

The third spectrum was obtained in August 2004 using SINFONI, a combination of SPIFFI and the adaptive optics (AO) system MACAO (Bonnet et al. 2003, 2004) at VLT-UT4. As SPIFFI was equipped with a new 2K×2K detector in early 2004, the data cubes know had a dimension of 64×32 pixels in the two spatial dimensions and 2048 pixels along the spectral axis. The data were processed and reduced like the earlier SPIFFI data. The pixel scale of the images was 100×50 mas/pixel, meaning that each image pixel is rectangular. Observations were obtained at a seeing of around 0.8", the spatial resolution was improved to 0.3" by the AO system. The spectrum covered the K band from 1.95 to 2.45μm with a spectral resolution of $R = 4500$.

We corrected each spectrum for background flux by subtracting a spectrum obtained in an empty field 0.9" north of IRS34W. To assure the comparability of the spectra and to estimate the influence of contamination by the neighbouring stars (especially by IRS34E located 0.4" north east of IRS34W), we obtained the 2003 SPIFFI spectrum twice: once on the original data cube, and once on a data cube in which (1) the spatial resolution (0.3") was reduced to 3D resolution (0.6") by convolving the spatial images with a gaussian, (2) the image pixels (pixel scale 100 mas/pixel) were combined to pixels of the size of 3D pixels (300 mas/pixel) and (3) the spectral resolution ($R = 3600$) was reduced to that of 3D ($R = 1500$) by convolving the spectral axis with a gaussian.

2.3 Results and discussion

The H and K band light curves and colour of IRS34W obtained from the stellar photometry are presented in Fig. 2.1. The magnitudes used in this discussion are not

2.3. Results and discussion

corrected for the (not well-known) extinction by interstellar matter, the given times are days after JD 2,448,000.

Both lightcurves show variations in magnitude on timescales between few 100 and around 1500 days, a selection of representative data points (one value per calendar year) is presented in Table 2.3. Especially the K lightcurve, showing an overall variation of $\Delta K \simeq 1.5$, does not show any periodicity or regularity, including periods of nearly constant magnitude as well as steplike changes in brightness.

The stellar colour $H - K$ shows variations which are clearly correlated to the lightcurves. From day 845 to day 3731 $H - K$ gets redder by about 0.8, whereas in the same time K decreases by 1.5. Within days 3731 and 4516 K increases again by around 0.4 and the colour gets bluer by roughly 0.8. We do not observe, within the errors, a significant variation in the time from day 4516 to day 5271, while in the same time span H and K both vary by around 0.5 magnitudes.

Variations on small timescales (few days) are not observed. Although the SHARP K lightcurve shows groups of data points not resolved in time (they were obtained in observing runs lasting 4–10 days) which appear to be somewhat scattered, these values are identical within their errors. One should note that the Gemini K point at day 3728 appears to be slightly offset compared to the SHARP points; but taking into account the error bars and the fact that these points were extracted from two different, seperately processed data sets, this is no significant indication towards short time variability either.

The normalized spectra of IRS34W obtained in 1996, 2003 and 2004 are presented in Fig. 2.2, all wavelengths used below are given in microns. All spectra show as prominent features the lines He I (λ 2.058), He I (λ 2.112), N III (λ 2.116), He I (λ 2.162), He I (λ 2.164) and H I (λ 2.166) in emission. Due to line blending the lines He I (λ 2.112) and N III (λ 2.116) on the one hand and He I (λ 2.162), He I (λ 2.164) and H I (λ 2.166) on the other hand are not seperated and are therefore here treated integrally as line complexes He I (λ 2.112)+N III (λ 2.116) resp. He I (λ 2.162)+He I (λ 2.164)+H I (λ 2.166). All lines resp. line complexes show P Cygni profiles identifying IRS34W as a wind source. We especially note a remarkable similarity between the spectra of IRS34W and the spectrum of the Ofpe/WN9 star HDE 269445 presented in Morris et al. (1996).

The comparison of the spectra shows only slight variations in the equivalent widths (EWs) of all emission lines (see Table 2.3) that are too small to be significant. A direct graphical comparison by overplotting the spectra (see Fig. 2.2, bottom panel) shows that they appear to be identical within their typical accuracies.

Obviously, the photometric and spectroscopic observations of IRS34W lead to am-

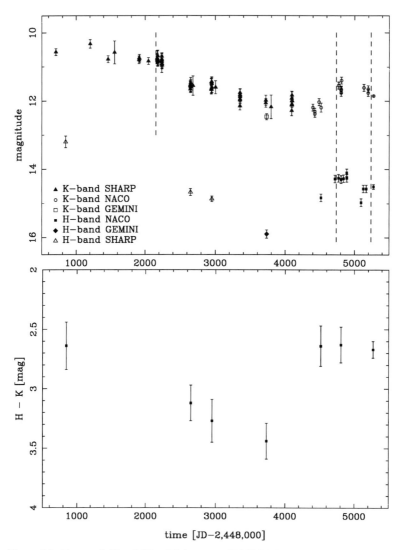

Figure 2.1: *Top panel:* H and K band lightcurves of IRS34W. The time is given as reduced Julian Date, the first data point was taken in March 1992, the last in September 2004. The vertical dashed lines indicate the times when the three spectra were obtained in 1996, 2003 and 2004. Strong non-periodic variabillity on time scales from months to years is clearly visible in both light curves. *Bottom panel:* H-K derived from the lightcurves. The comparison to the lightcurves shows that the colour gets redder when the magnitudes decrease, and bluer when the magnitudes increase.

2.3. Results and discussion

time	H	K	$H-K$
705		10.56±0.10	
			2.64±0.20[a]
845	13.20±0.17		
1203		10.33±0.12	
1465		10.77±0.10	
1917		10.75±0.12	
2241		10.81±0.11	
2649	14.67±0.10	11.55±0.11	3.12±0.15
2949	14.87±0.08	11.60±0.16	3.27±0.18
3350		11.95±0.13	
3728		12.46±0.09	
			3.44±0.15[b]
3731	15.90±0.12		
4094		12.09±0.10	
4516	14.83±0.11	12.19±0.13	2.64±0.17
4806	14.29±0.11	11.66±0.10	2.63±0.15
5271	14.51±0.07	11.85±0.02	2.67±0.07

[a] This colour is computed from the magnitudes at days 705 and 845.

[b] This value is computed from the magnitudes at days 3728 and 3731. Both H and K were extracted from the Gemini North data set.

Table 2.1: Selected H and K magnitudes (one data point per calendar year) and colours of IRS34W, the values are not corrected for extinction. The time is given in days after JD 2,448,000. A months- to years-time-scale variability can be seen in H and K, including variations of the stellar colour.

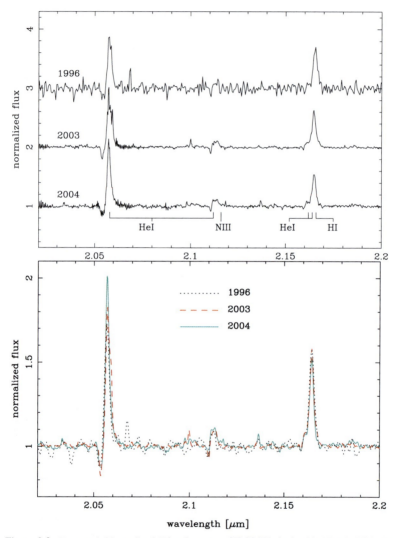

Figure 2.2: *Top panel:* Normalized *K* band spectra of IRS34W obtained in March 1996 with 3D, in April 2003 with SPIFFI and in August 2004 with SINFONI (spectra shifted along the flux axis). *Bottom panel:* The same spectra, smoothed and overplotted. Obviously, the spectra are identical within their typical accuracies.

2.3. Results and discussion

	A	B	C
1996	22.1±1.9	3.2±1.6	18.7±1.9
2003[a]	26.0±0.6	5.4±0.7	18.5±0.6
2003[b]	25.4±0.9	3.5±0.8	18.5±0.5
2004	25.8±0.8	4.4±0.9	15.6±0.9

[a] Spectrum taken from SPIFFI cube reduced to pixel scale, spatial and spectral resolution of 3D data.

[b] Spectrum obtained from original SPIFFI cube.

Table 2.2: Equivalent widths of the dominant emission lines in the March 1996, April 2003 and August 2004 spectra. Column "A" is He I (λ 2.058), "B" He I (λ 2.112) +N III (λ 2.116) and "C" He I (λ 2.162)+He I (λ 2.164)+H I (λ 2.166). The wavelengths of the lines are given in microns, the equivalent widths are given in Å.

bivalent results. On the one hand, the photometric magnitude of the star is highly variable on timescales from months to years. The shape of the light curve and the time scales of the variability rule out the possibility that IRS34W is an eclipsing binary. The amplitude of the variation, up to $\simeq$ 75% in flux, shows that an eclipsing star would have to cover at least that fraction of IRS34W's surface on the sky. The lack of significant variation in the line to continuum ratios shows that the continuum of this eclipsing star would have to be small relative to the fraction of the flux of IRS34W that remains to be seen. In that case, the temperature of the eclipsing star would be below $\simeq$5000K, and it would have CO absorption lines, that we would be able to detect even diluted in the continuum of the blue supergiant. Therefore also a coincidental eclipse by an unrelated star can be excluded.

Another possible effect to be examined is the obscuration of IRS34W by interstellar matter. The central parsec is known to contain several gas patches of various dimensions, that can be responsible for local enhancements of the extinction by $\Delta K \simeq 1$ (Paumard et al. 2004), and could therefore in principle account for the variability of IRS34W. Such clouds are however several arcseconds wide, and should therefore effect not only IRS34W but also neighbouring stars. Indeed, the lightcurves of some faint stars ($K \simeq 16$) taken from the original automatically generated NACO photometric data set (Trippe 2004) in a distance up to 0.9" from IRS34W showed similar, but weaker variations. Therefore we reanalyzed the magnitudes of IRS34W and its neighbouring stars in a set of selected good NACO images, applying additional background flux subtraction and Lucy-Richardson deconvolution instead of Wiener filtering. This analysis confirmed the magnitudes and variations found for IRS34W and made it possible to trace irregularities in some faint stars back to flux spill-over effects from IRS34W to stars located in its seeing halo. This rules out as well a cloud in the line of sight but much closer to the Earth, as it would be even larger in projection, and cause similar variability in a large number of stars, that would be obvious in our data. So we finally could conclude, that the photometric variability is indeed bound to the star resp. its immediate vicinity (in terms of the given spatial resolution) and not to an external source.

On the other hand, the absence of significant spectroscopic variability complicates the understanding of the stellar behaviour. As the emission lines we observe in our spectra are sensitive especially to the stellar parameters T_{eff} and $\dot{M}$, significant variations of these parameters should be visible in the spectra. An example for this is the LBV AFGL 2298 that shows photometric variations in K band of around 0.3 magnitudes and variations in the EWs of He I (λ 2.058) and H I (λ 2.166) by factors of 3 to 4 within one year (Clark et al. 2003a). Indeed, our observations indicate that at least T_{eff} and $\dot{M}$ remain rather constant while the star gets dimmer or brighter.

Given that we do not see significant changes in the spectral lines, variations of the wind properties are highly improbable since this would mean a change of the density (by modifications either of maximum velocity at the top of the wind, v_{∞}, or $\dot{M}$), which would lead to significant observable spectroscopic variations.

Taking this into account, we find two possible interpretations for the photometric variability of IRS34W:

(1) The variability is connected to the physics of the stellar photosphere. Since T_{eff} did not change, this assumption would lead to the conclusion that the stellar radius (as $L \propto R_*^2 \cdot T_{\text{eff}}^4$) changed by a factor of around 2. Indeed Dorfi & Gautschy (2000) observe comparable variations in stellar radii (by a factor up to $\simeq 1.6$) in their simulations of pulsations of massive stars, but on timescales of days, not years as in the case of IRS34W. More interestingly, they also note that the *raise* of a pulsating mode can be accompanied by a change in stellar radius (a factor of $\simeq 1.3$ in their model) and only a little increase in T_{eff} (2700K on top of 18000K, thus probably not observable for us), on the timescale of two years. Although in this particular case the overall change in luminosity is smaller than the variations we observe ($\Delta K \simeq 0.44$ as calculated from a black-body distribution for the given changes in temperature and radius), it makes it plausible that we may have witnessed a phase transition between a hydrostatic equilibrium and a pulsating mode.

This explanation is complicated by the fact that a change in stellar radius would change the observed EWs, which scale as $\dot{M}/R_*^{1.5}$. Thus the constancy of the EWs would require a corresponding increase of $\dot{M}$ parallel to the rise of the radius.

2.3. Results and discussion

Apart from this rather improbable scenario, it is hard to understand how a significant change in stellar radius, accompanied by a new hydrostatic quasi-equilibrium that would last for years, would not be accompanied by a significant change in T_{eff}.

(2) The photometric variations of IRS34W are not caused by the photosphere but from above the wind. The colour of the star, significantly redder than other stars of similar brightness (e.g. our photometric reference star ID185 located 3" NE of IRS34W with $K = 12.09 \pm 0.08$ and $H - K = 2.26 \pm 0.13$) in the field, also suggests that it is affected by a higher extinction.

This is supported by Clenet et al. (2001) who derive a colour of $K - L = 2.4$ which is about 1 magnitude redder than the colours of other He I resp. Ofpe/WN9 stars observed in the field. Additionally, Viehmann et al. (2005) find the $H - K$ colours of other Galactic Centre He I stars to be clearly bluer than what we find for IRS34W.

As we were able to exclude external effects like stars or interstellar clouds, the variability would then be due to variations in the column density of circumstellar material expelled by the star.

Both cases - from which scenario (2) is obviously the more probable one - suggest the identification of IRS34W as a LBV - or close to this stage. It must be noted that a slightly elongated nebular emission feature is visible around IRS34W on a $H/K/L$ NACO colour map that traces the dust emission (Genzel et al. 2003a, Fig. 1 right), and that this feature could as well be an interstellar structure as circumstellar material from IRS34W that might have been ejected in earlier LBV type eruptions. It is also interesting to note that the shape of the K band light curve of IRS34W is very similar to the V light curve of the LBV AG Car (see Humphreys & Davidson 1994): a sudden decrease is followed by a plateau and then a slight increase in a time span of around six years. But contrary to IRS34W, in case of AG Car the photometric variability is accompanied by strong spectroscopic variability.

Another interesting point supporting the "dust scenario" is given by Clark et al. (2003b). Indeed, their mid-IR study of two ring nebulae shows that they contain stars with properties very similar to candidate LBVs. The K band spectra clearly identify them as luminous evolved massive stars. They are also similar to the spectrum of IRS34W, being dominated by H and He I emission lines. The surrounding nebulae are explained by direct ejection of material by the star and/or by interaction of ejected material with the interstellar medium. Among the two stars, G24.73+0.69 is especially interesting since it is photometrially variable ($\Delta K \simeq 0.3$) over $\simeq 1.5$ years, but shows only little spectroscopic variability (see Fig. 10 of Clark et al. 2003b). This similarity favors the explanation of the observed behaviour of IRS34W by the presence of circumstellar dust.

In addition, Clark et al. (2005) included IRS34W in their list of confirmed LBVs, although this is partly based on the assumed variation of stellar and wind parameters that we show is not likely. However, they mention the possibility of a dusty circumstellar environment to explain the magnitude variations of IRS34W, in agreement with the present study. They note that only LBVs, B[e] supergiants and late WC stars are known to be dust producers. If dust is indeed the explanation for the variability of IRS34W, this is then an indication that it is closely related to Luminous Blue Variables, despite the lack of variation in stellar properties.

Figure 2.3 shows the approximated position of IRS34W in a Hertzsprung-Russell diagram. The luminosity was estimated as follows: A_K was derived from the V band extinction map of Scoville et al. (2003) and the extinction law of Moneti et al. (2001), and was used together with the observed K magnitude to estimate M_K. Then, bolometric corrections computed from our preliminary modelling of Ofpe/WN9 stars covering a wide range of physical parameters (L, T_{eff}, $\dot{M}$) were used to finally derive L. The extension of the area takes into account uncertainties in the extinction and the bolometric corrections. As for the effective temperature, we only show a range of values for which, within the L range, He I (λ 2.058) shows a P-Cygni profile in our preliminary models. So far this is a rather rough estimate, but gives some interesting indications on the evolutionary status of this star. Compared to already known Ofpe/WN9 stars, LBVs and O supergiants, IRS34W is located between the O supergiant and LBV stages (like other known Ofpe/WN9 stars, see Fig. 2.3) and therefore appears to be an object in transition between these two phases.

The estimate of the luminosity of IRS34W presented above shows that it may be intrinsically less luminous than the other He I stars (see Najarro et al. 1997). As discussed above, it is, regardless of the variations reported here, also redder. Given this and the fact that dust formation is the most likely explanation for the observed variability, we can imagine the following scenario: due to its lower luminosity, IRS34W is able to experience dust formation in its atmosphere. Such episodes of dust formation makes the star fainter and redder; at the end of each episode, dust is destroyed, but only partially, leading to a long-term accumulation of dust around the star so that on average, IRS34W is redder (and also even fainter) than the other He I stars. The key point of this scenario is the lower intrinsic luminosity of the star previous to any dust formation episode.

This is very speculative however since the formation of dust in atmospheres of hot stars is still poorly understood, although Cherchneff & Tielens (1995) argue that special geometries such as equatorial disks can lead to the high densities necessary for dust formation; other properties such as a thick wind or inhomogeneities in the

2.3. Results and discussion

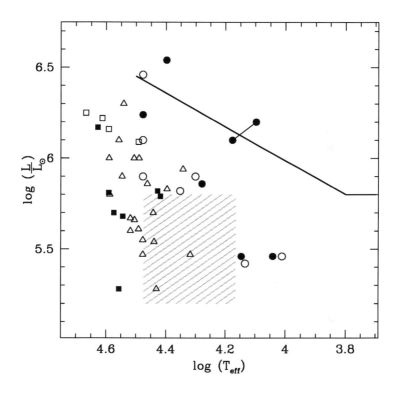

Figure 2.3: The approximated position of IRS34W (shaded area) in a HRD compared to already known similar stars. Triangles represent Ofpe/WN9 stars (Pasquali et al. 1997, Crowther et al. 1995, Bianchi et al. 2004, Bresolin et al. 2002), circles LBVs (Humphreys & Davidson 1994, Clark et al. 2003a) and squares O supergiants (Bohannan & Crowther 1999, Repolust et al. 2004, Crowther et al. 2002), filled symbols mark Galactic stars, empty symbols are stars in the Large Magellanic Cloud and NGC 300; the solid line is the Humphreys-Davidson limit. This diagram shows that IRS34W is located between O supergiants and LBVs.

atmosphere may also be necessary. In the other He I stars of the Galactic Center, the luminosity may be too high, so that dust formation is inhibited by the too strong radiation. In IRS34W, the photon flux may be low enough to allow dust formation.

2.4 Summary

In this article we reported the results of time-resolved photometry and spectroscopy of the Galactic Centre star IRS34W and possible interpretations, which can be summarized as follows:

1. The photometric magnitude of IRS34W is highly variable on timescales from months to years with an overall variation of $\Delta K \simeq 1.5$.

2. Variations of the stellar colour $H - K$ track the photometric variability. When the brightness decreases, the colour gets redder, when the brightness increases, the colour gets bluer.

3. The spectra of the star do not show any significant variations over a time span of 8 years.

4. The most probable reason for the stellar behaviour is an obscuration by circumstellar material ejected by the star.

5. IRS34W is a star in transition between the O supergiant and LBV phases, only the lack of spectroscopic variability prevents us from safely identifying it as a LBV.

IRS34W and five additional stars, which are very similar to IRS34W, have been classified as LBV candidates by Paumard et al. (2001). Among these additional stars only one, IRS16SW, shows strong, periodic variability (Ott et al. 1999). Spectrophotometric monitoring of these stars is currently under way; detailed quantitative analysis and modelling of this group of stars, including IRS34W, should shed more light on the physics of this class of objects.

Acknowledgements: We thank the referee, F. Najarro, for helpful suggestions and comments that helped to improve the quality of the paper. F.M. acknowledges support from the Alexander von Humboldt Foundation.

Chapter 3

Polarised emission from SgrA*

Original publication: S. Trippe, T. Paumard, T. Ott, S. Gillessen, F. Eisenhauer, F. Martins & R. Genzel 2007, *A polarised infrared flare from Sagittarius A* and the signatures of orbiting plasma hotspots*, MNRAS, 375, 764

Abstract: In this article we summarise and discuss the infrared, radio, and X-ray emission from the supermassive black hole in the Galactic Centre, SgrA*. We include new results from near-infrared polarimetric imaging observations obtained on May 31st, 2006. In that night, a strong flare in K_s band (2.08 µm) reaching top fluxes of ∼16 mJy could be observed. This flare was highly polarised (up to ∼40 %) and showed clear sub-structure on a time scale of 15 minutes, including a swing in the polarisation angle of about 70 degrees. For the first time we were able to observe both polarised flux and short-time variability, with high significance in the same flare event. This result adds decisive information to the puzzle of the SgrA* activity. The observed polarisation angle during the flare peak is the same as observed in two events in 2004 and 2005. Our observations strongly support the dynamical emission model of a decaying plasma hotspot orbiting SgrA* on a relativistic orbit. The observed polarisation parameters and their variability with time might allow to constrain the orientation of accretion disc and spin axis with respect to the Galaxy.

3.1 Introduction

The centre of our Milky Way hosts the 3.6-million-$M_\odot$ supermassive black hole and radio source SgrA*. This black hole is generally invisible in NIR wavelengths and was not detected in this spectral range before 2002 when diffraction-limited observations at 8-m-class telescopes became possible (Genzel et al. 2003b; Ghez et al. 2004).

Since then, several NIR flares, which appear on time scales of few events per day, have been observed photometrically (Ghez et al. 2005b; Eckart et al. 2006a), spectroscopically (Eisenhauer et al. 2005a), and polarimetrically (Eckart et al. 2006b). Such flares last typically for about 60–120 minutes. These observations gave information on the colours and spectral indices of flares (Ghez et al. 2005b; Eisenhauer et al. 2005a; Gillessen et al. 2006; Krabbe et al. 2006). They included the detection of polarised flux and quasi-periodic substructures on time scales of 15 to 20 minutes (Genzel et al. 2003b; Eckart et al. 2006b).

In recent years, variable and flaring emission from SgrA* has been observed in a variety of wavelength bands, especially in the radio (Aitken et al. 2000; Melia & Falcke 2001 [and references therein]; Bower et al. 1999a,b, 2003a ; Miyazaki et al. 2004; Marrone et al. 2006) and X-ray (Baganoff et al. 2001, 2003; Goldwurm et al. 2003; Aschenbach et al. 2004; Bélanger et al. 2005, 2006) regimes. Sub-structure on minute time scales was also detected in X-ray flares (Aschenbach et al. 2004; Bélanger et al. 2006). Additionally, variable polarised flux from SgrA* was found in sub-mm to mm radio bands (Bower et al. 205; Marrone et al. 2006; Macquart et al. 2006).

In this article we discuss the physics behind the emission from SgrA* taking into account new results of polarimetric imaging observations obtained in May 2006. In section 2 we describe the observations and the data reduction, in section 3 we present the observational results. In section 4 these data are placed into the context of earlier results, and in section 5 they are discussed and interpreted.

3.2 Observations and data reduction

We have repeatedly carried out observations on the 8.2-m-UT4 (Yepun) of the ESO-VLT on Cerro Paranal, Chile, using the detector system NAOS/CONICA (NACO for short) consisting of the AO system NAOS (Rousset et al. 2003) and the 1024×1024-pixel NIR camera CONICA (Lenzen et al. 2003).

On May 31st, 2006, in total 240 minutes of polarimetric K_s band ($\lambda_{center} = 2.08 \mu m$) imaging data of the Galactic Centre were obtained. The Wollaston prism mode of NACO made it possible to simultaneously observe two orthogonal polarisation angles (corresponding to the ordinary and the extraordinary beam of the prism respectively) per image. In order to cover a sufficient number of polarimetric channels, the observed angles were switched using a half-wave retarder plate.

The images were obtained alternately covering the polarisation angles 0°/90° and 45°/135° respectively. Each cycle took no more than about 150 seconds. The spatial resolution of the data is around 60 mas at mediocre Strehl ratios. All frames have a pixel scale of 13.27 mas/pixel.

3.2. Observations and data reduction

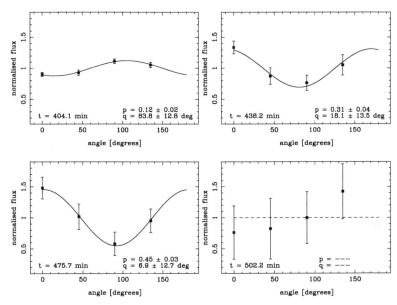

Figure 3.1: Four examples for the calculation of degrees and angles of polarisation as presented in Figs. 3.3 and 3.5. Each panel shows the normalised and calibrated flux of SgrA* in the four polarimetric channels belonging to a given time bin. The four flux values are plotted together with the best fitting sine curve used to compute the degree of polarisation p and the polarisation angle q at time t. In each case, p is given by the amplitude of the sine and q by the phase. In cases where the flux values are identical within the errors (bottom right panel), no polarisation parameters are calculated.

All images were sky subtracted, bad-pixel- and flat-field-corrected. To extract the fluxes of SgrA* and two comparison stars, we applied aperture photometry. 8 bright and isolated stars in the field of view served as calibrator sources. As SgrA* was confused with a weak star, S17, at the observation epoch, this star's flux contribution of 2.5 mJy was subtracted from the flare data.

As always two pairs of polarisation channels were observed alternately, for each source sets of four flux values (for the four angles) per time bin were obtained. In order to extract the polarimetric parameters – degree of polarisation p, angle of polarisation q – from a given data set, this data set was first normalised by dividing all four values by their average. Due to this, the average of the data set is set to 1. The amplitude of variations *around the average level* due to polarised flux is limited to the range from 0 to 1. As we use the convention that the degree of polarisation is the ratio of polarised flux vs. total flux, this amplitude corresponds to the degree of polarisa-

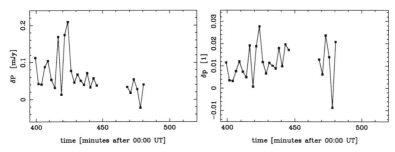

Figure 3.2: Comparison of two methods for calculating polarisation parameters. Polarised flux P and polarisation fraction p were computed (a) numerically via a sine fitting method, (b) analytically using the Stokes parameters I, Q, and U. The analytical approach was explicitely corrected for bias. Both panels show the differences between the two approaches; this allows to estimate the influence of bias. *Left panel*: Differences in polarised flux. The average deviation betwen the two methods is 0.064±0.073 mJy. *Right panel*: Differences in polarisation fraction. The average deviation is 0.012±0.013. As all differences are zero within the quoted errors (see also Figs. 3.3, 3.5), the bias is not significant.

tion. We compute the polarised flux as the product of the degree of polarisation and the total source flux. One should note that throughout this article "polarisation" and "polarimetry" refer to *linear* polarisation.

As described above, in each image source fluxes are calibrated by dividing the target source count rates by the count rates of calibration stars taken from the same image (photometric calibration). This means that a possible average polarisation of the calibrator stars is erased. As indeed the stars of the observed Galactic Centre region show an average polarisation of 4% at an angle of 25° due to foreground extinction by dust (Eckart et al. 1995; Ott et al. 1999), this has to be corrected. This correction was done by multiplying the four flux values of a data set with the function

$$f(\phi) = 1 + 0.04 \cdot \sin(2 \cdot (\phi + 25°)) , \quad \phi = 0°, 45°, 90°, 135° \quad (3.1)$$

(polarimetric calibration). The factor 2 in the argument of the sine is due to the convention that polarisation angles are limited to the range [0°, 180°]. After this, each normalised data set was fit with a sine curve with a period of 180°. This delivers the degree of polarisation (the sine curve's amplitude) and the angle of polarisation (the sine curve's phase).

In our definition, a polarisation angle of 0° corresponds to a pointing to the north and the angle is counted east of north. In both the polarimetric calibration and the fitting, a global rotation of the polarisation vector with respect to the sky of 36° was taken

3.2. Observations and data reduction

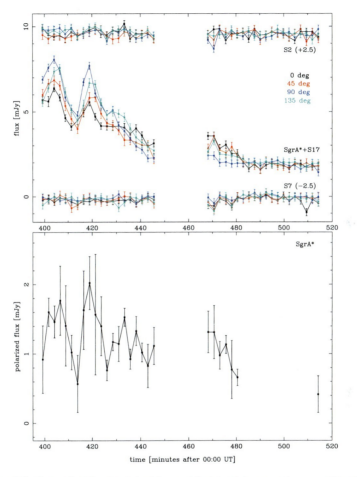

Figure 3.3: Observed total and polarised fluxes of SgrA* and the comparison stars S2 and S7. *Top panel:* Lightcurves of SgrA*, S2, and S7 separated into the four polarimetric channels. Due to the splitting of the light into one ordinary and one extraordinary beam, each channel in average contains one half of the total source flux. These values are the observed fluxes, before calibrating relative to the extinction screen polarisation and subtraction of the confusing star S17. S2 and S7 are shifted along the flux axis. The fast variability of SgrA* and its strong polarisation can be seen clearly, especially in comparison to S2 and S7. *Bottom panel:* polarised flux of SgrA*. The two main maxima in the polarised flux correspond to the maxima in the lightcurve. In all figures the central gap in the data is due to sky observations, around a time $t = 480$ min the flare fades out. Where data points are missing, no parameters were computed (see Sect. 2 for details).

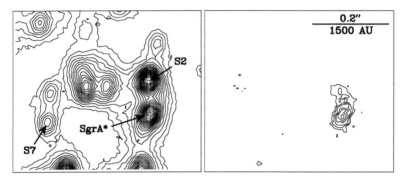

Figure 3.4: Contour maps of integrated and polarised emission from SgrA*; N is up, E to the left. *Left:* Sum image of the channels 0° and 90° at the time of the first flux peak ($t = 405$) showing the immediate vicinity of SgrA*. Contours are 20, 25, ... 120 times noise level. *Right:* Difference of the channel maps used for the left image. Contours are 4, 6, ... 16 times noise level. The strong residual source at the position of the flare – corresponding to the polarised flux – is clearly visible.

into account. This rotation was found using polarimetric NACO Wollaston data of the calibrator star IRS21 obtained in July 2005. Comparing the parameters extracted from this data set with results found with different instruments and reported earlier (14% and 14°; Eckart et al. 1995, Ott et al. 1999) leads to a rotation of 34°. This rotation might be caused by a known shift (the abovementioned 36°) in the zero position of the half-wave plate, although this shift is assumed to be corrected in the instrument setup (N. Ageorges, *priv. comm.*; NACO HWP commissioning report). Given the numerical agreement (better than 2°) we believe that at least in this data set a correction was not applied.

The sine fitting procedure described above is demonstrated in Fig. 3.1 for four different time bins. Here the connection between p and q on the one side and amplitude and phase of the flux data on the other side is obvious. This figure already indicates some evolution of the polarisation parameters with time; details will be discussed below.

Data sets, for which the values were consistent with being identical within the errors (corresponding to a $\chi^2_{\text{reduced}} < 0.789$ in case of three degrees of freedom) were not fit in order not to apply a systematically incorrect model. Another effect to be taken into account was the bias caused by the non-zero errors of the flux values. Bias can lead to a systematic overestimation of polarisation fraction and polarised flux especially in cases of low fluxes. In order to check this, we calculated p and polarised flux using analytical expressions for the Stokes parameters I, Q, and U corrected for the bias terms. The results are presented in Fig. 3.2. In no case the deviation exceeded 3% (in p; average deviation: 1.15±1.34%) resp. 0.2 mJy (polarised flux; average deviation: 0.064±0.073 mJy). As all deviations are zero within the quoted errors, the influence of bias can be safely neglected for the further discussion.

3.3 Results

The lightcurves for SgrA* and two comparison stars, S2 and S7, are presented in Fig. 3.3 (top panel). In this figure the lightcurves are shown for each polarisation channel separately. The presented values are the observed fluxes before polarimetric calibration relative to the foreground extinction screen and before subtracting the flux of S17. In all lightcurves gaps due to intermediate sky observations are present. All times mentioned here and elsewhere in this article are minutes after 00:00 UT of the observation day, the first data point obtained is located at $t = 399$ min.

From these lightcurves one can see a strong and fast variability of the flare. Especially interesting is its double peak, which shows flux variations in the order of 7 mJy or 40% within 10 minutes time. The two maxima of the double peak are separate in time by only 15 minutes. Additionally, the different fluxes in the polarimetric channels point towards significant polarisation. This is demonstrated more directly in Fig. 3.4, where the difference of two channels is mapped and compared to a sum image of the vicinity of SgrA*. It is important to note that those differences in the polarimetric channels are not visible in the lightcurves of the comparison stars. This holds for both flux levels corresponding to the peaks of the flare (S2) and for fluxes corresponding to the end of the flare (S7).

The amount of observed polarised flux from SgrA* is shown in the bottom panel of Fig. 3.3. The polarised flux shows two maxima at times 405 resp. 420 reaching up to around 2 mJy; these maxima correspond to the double peak in the lightcurves. During the entire flare the polarised flux is never lower than $\sim$0.6 mJy.

The evolution of the parameters degree and angle of polarisation is presented in Fig. 3.5. The degree of polarisation is about 15% at the beginning of the observation and increases after the second main maximum ($t > 430$) up to roughly 40%. This can be seen also in Fig. 3.1, were two examples for the fluxes in the four polarimetric channels are given for time bins corresponding to the first peak ($t = 404$) resp. the end of the flare ($t = 476$). The increase is a direct consequence of the fact that the polarised flux level remains roughly constant while the overall flux is decreasing with time after the double peak.

Inspecting Fig. 3.3 and 3.5, it might not be obvious that for times $t < 430$ the

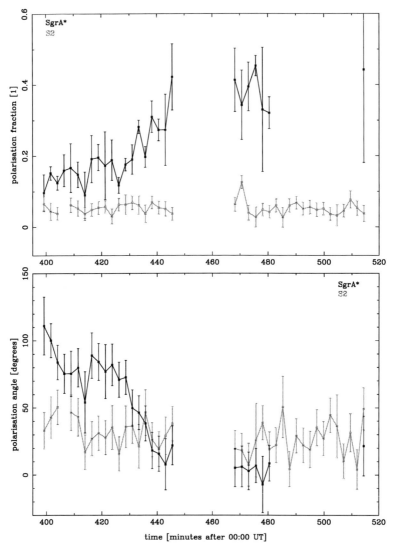

Figure 3.5: Evolution of degree and angle of polarisation for SgrA* and S2 during the flare. *Top panel:* Degree of polarisation. During the flare ($t < 480$) it never sinks below 10%, reaching top levels of $\sim$40% . *Bottom panel:* Angle of polarisation. Of special interest is the strong swing of $\sim$70° occuring in the time range 430...445 min, i.e. within 15 minutes. In both panels the results for S2 mirror the calibration. Where data points are missing, no parameters were computed (see Sect. 2 for details).

3.3. Results

polarised flux is variable and the degree of polarisation is not. To check this, we computed the reduced χ^2 for each data set using the assumption that the data do not vary with time. We find a $\chi^2_{\text{reduced}} = 2.12$ for the polarised flux, whereas for the degree of polarisation we find $\chi^2_{\text{reduced}} = 0.88$. Thus we conclude that indeed only the polarised flux varies for $t < 430$, whereas the degree of polarisation remains constant within the errors.

The polarisation angle remains at a constant level around 80° during the time of the double peak, i.e. $t \leq 430$. The first two data points indicate that the angle could have been even larger before the first peak; but as this signal is only marginally significant (1–1.5σ), this is completely speculative. Beginning at $t \sim 430$, q swings by about 70° within 15 minutes, reaching values down to ~10°. This is (within the errors) close to the angle of the foreground polarisation (~25°).

When discussing the angle of polarisation, one has to keep in mind possible calibration artefacts. As described in Sect. 2, our data are calibrated so that an intrinsically unpolarised source shows a signal corresponding to $p = 4\%$ and $q = 25°$. Therefore it is in general possible to observe a source composed of two superposed flux components, a polarised one, and a non-polarised one. As long as both components are significantly bright, one would measure the polarisation parameters of the polarised source flux. But when the polarised component fades away, the observed p and q would move towards the foreground level.

To assure that we are not mislead by such an effect, the comparison to a source with a brightness similar to SgrA* (plus S17) at the very end of the flare ($t = 470...480$) without applying any polarimetric calibration becomes important. Indeed this is what is shown in Fig. 3.3, where the photometric lightcurves of SgrA* and S7 are compared. As one can see, even at the very end of the flare SgrA* is clearly polarised intrinsically, whereas S7 is not at all. Additionally, we repeated the fitting of the polarisation parameters without re-introducing the foreground polarisation into the data. The results turned out to be identical within the erors compared to those shown in Figs. 3.3 and 3.5. Thus we are confident that the description above is valid.

Morphologically, the flare shows two phases. In the first phase, covering times $t \leq 430$, the double peak occurs, the polarised flux changes rapidly and traces the overall emission, and both degree and angle of polarisation remain constant.

In the second phase ($t > 430$) the overall flare slowly fades away while the polarised flux remains on a roughly constant level, leading to an increase in the degree of polarisation. Additionally the swing in polarisation angle occurs.

Following the evolution of the flare with time, one has to note that it is highly dynamic on a typical time scale of 15 minutes, which expresses itself in all parameters: the overall flux (double peak), polarised flux, polarisation angle, and degree of polarisation.

3.4 Context

The SgrA* activity described above for the first time combines several separately observed properties on high significance levels: (1) strong, variable NIR activity with an overall duration of more than 80 minutes, (2) sub-structure on a time scale of 15 minutes, and (3) clear, variable polarisation.

These signatures allow a deeper understanding of the physical emission mechanisms of SgrA*. This is especially obvious in the long-term context of observed infrared, radio, and X-ray activity. For this reason the overall properties of SgrA* flares are discussed and compared here.

Since 2002 we have observed the NIR emission of SgrA* using the ESO Very Large Telescope on Cerro Paranal, Chile. Photometric and polarimetric imaging data were collected with NACO in H, K, and L bands (1.3 – 4.1 μm; Genzel et al. 2003b, Trippe 2004, Eckart et al. 2006a,b).

Infrared spectra were obtained using SINFONI, a combination of the integral field spectrometer SPIFFI (Eisenhauer et al. 2003b,c) and the adaptive optics (AO) system MACAO (Bonnet et al. 2003, 2004), at VLT-UT4. The data covered the K band from 1.95 to 2.45 μm with a spectral resolution of $R = 4500$ (Eisenhauer et al. 2005a; Gillessen et al. 2006). The properties of 16 observed infrared flares are given in Table 3.4.

Additional to our work, infrared observations of SgrA* were obtained by Ghez et al. (2004, 2005b) in K, L, and M bands using the Keck II telescope on Hawaii. Emission in L and M bands observed with NACO was also reported by Clénet et al. (2004, 2005). Yusef-Zadeh et al. (2006a) used the Hubble Space Telescope to monitor flares in the range 1.60–1.90μm. Recently, Krabbe et al. (2006) collected spectroimaging data of a K band flare using the integral field spectrometer OSIRIS at the Keck II telescope.

In radio wavelengths (sub-mm to cm) SgrA* was discovered by Balick & Brown (1974) and has since then been monitored extensively (recently e.g. Aitken et al. 2000; Melia & Falcke 2001; Bower et al. 1999a,b, 2003a, 2005; Miyazaki et al. 2004; Marrone et al. 2006; Macquart et al. 2006) photometrically, spectroscopically, and polarimetrically using a large ensemble of telescope facilities.

In X-ray wavelengths (few keV) SgrA* flares have been observed photometrically and spectroscopically since 1999 using the NASA Chandra and the ESA XMM-

3.4. Context   71

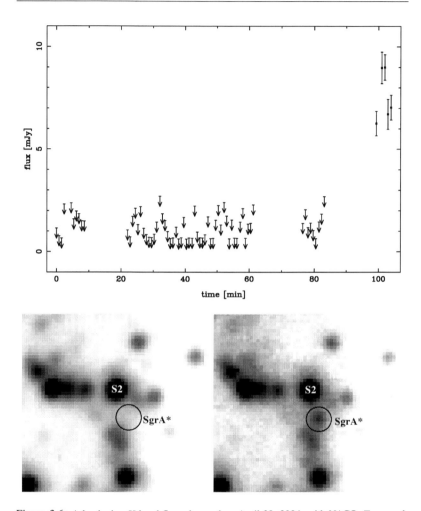

Figure 3.6: A beginning *H* band flare observed on April 28, 2004, with NACO. *Top panel*: Lightcurve of the event. The flare begins after $t \approx 85$ min. Before this time, no flux is detected at the position of SgrA*. For $t < 85$ min the upper limits for source flux are given. Gaps are due to sky observations and a short breakdown of the AO system. *Bottom panels*: Images showing SgrA* before and after beginning of the flare. The *left hand* image is an average of 20 frames obtained in the time range $t = 45...60$ min. The position of SgrA* is free of any excess emission. The *right hand* image is an average of the last five frames obtained. Here SgrA* is clearly visible. This example data set illustrates the flaring character of SgrA*: the black hole does not show any detectable activity for at least 1.5 hours, then a flare raises from zero level within minutes.

no.	epoch (years)	mode of observation	band	peak flux (mJy)	duration of flare (min)	time scale of sub-structure (min)	α (1)	p (%)	q (°)
1	2002.66	phot	L'	30	>15	–	–	–	–
2	2003.35	phot	H	16	30	–	–	–	–
3	2003.45	phot	K_s	13	80	20	–	–	–
4	2003.45	phot	K_s	9	85	17	–	–	–
5	2004.32	phot	H	9	>15	–	–	–	–
6	2004.45	pol	K_s	5	35	–	–	20	80
7	2004.51	phot	K_s	8	>250?[a]	25	–	–	–
8	2004.52	phot	K_s	3	85	13	–	–	–
9	2004.54	spec	K	3	60	–	-2.2	–	–
10	2004.54	spec	K	3	60	–	-3.5	–	–
11	2005.27	phot	K_s	3	>20	–	–	–	–
12	2005.46	spec	K	8	>150	20	-3...+2[b]	–	–
13	2005.57	pol	K_s	8	100	20	–	15	75
14	2006.40	phot	L'	25	110	–	–	–	–
15	2006.41	pol	K_s	16	>80	15	–	15...40[c]	80...10[c]
16	2006.42	phot	L'	23	>150?[a]	–	–	–	–

[a] Very uncertain due to poor data quality.

[b] Variations within the same flare and correlation with source flux.

[c] Variations within the same flare with time.

Table 3.1: Properties of infrared flares observed since 2002. Observations were obtained in photometric ("phot"), polarimetric ("pol"), and spectroscopic ("spec") modes. α is the colour index (defined as $\nu L_\nu \propto \nu^\alpha$), p the degree of polarisation, and q the angle of polarisation. Parameters marked "–" were not measured. Typical uncertainties are for fluxes: 1 mJy (L' band: 3 mJy); times: 2 min; α: 1; p: 5%; q: 10°.

Newton space telescopes (Baganoff et al. 2001, 2003; Goldwurm et al. 2003; Aschenbach et al. 2004; Bélanger et al. 2005, 2006; Eckart et al. 2006a).

This view over many seperate observations in several wavelength regimes allows some general statements:

- SgrA* emission is *flaring*. In NIR and X-ray wavelengths it is regularly detected in form of outbursts. In both bands flares correspond to an increase of flux by factors up to ∼10 from the background level within some ten minutes. The typical length of a flare is in the range of 1–3 hours. The flare event rate (i.e. the number of flares per time) is in the order of few events per day. For the 16 cases listed in Table 3.4 the flare rate is 2.5 events per day; including some flares covered by poor quality data increases this number to about ∼3.3 NIR events per day. In four cases NIR and X-ray flares were observed to be simulaneous within the available time resolutions (few minutes). Inspecting Table 3.4 shows

3.4. Context

(within the limits of low-number statistics) a general trend: flares are the more seldom, the more luminous they are. In contrast to this, changes in the radio flux are limited to variations of <50% within hours to days.

The flaring character of SgrA* is illustrated by the lightcurves presented in Fig. 3.3. Here the emission drops down to (and remains) zero (within the errors) in both total flux and polarised flux, after a phase of strong activity. Another example is given in Fig. 3.6. This figure presents an H band flare observed in April 2004. In this case, after more than one hour of zero emission from the position of the black hole, strong emission raises up to 9 mJy within about 20 minutes. In both cases the observations are inconsistent with a permanent, variable NIR source. An equivalent behaviour could be observed at several other occasions in both NIR and X-ray bands. Thus the classification of SgrA* emission as "flaring" is justified.

- SgrA* emission is *polarised*. Linear polarisation in the order of few to few ten per cent was detected in radio and NIR bands. In NIR, this polarisation is observed in flares as described in sections 2 and 3. For three NIR flares so far observed polarimetrically (flares 6, 13, and 15 in Table 3.4) we found polarisation degrees of 15–20% and angles of $\sim 80°$ on sky at times of maximum fluxes. The polarisation fractions of flares 6 and 13 do not show significant variations with time. In contrast to the flare described in Sect. 3, there are no distinct peak/decay phases. Unfortunately, these statements are weakened by the larger relative errors caused by lower peak fluxes (5 and 8 mJy in contrast to 16 mJy) of the flares 6 and 13.

Concerning the observed polarisation angle, it is important to note that this angle was found repeatedly in three measurements covering a time span of two years. This strongly suggests that the geometry of the emission region is stable in time.

In comparison to this, the continuous radio flux was found to be polarised with $p \approx 2$–8% and $q \approx 135$–165° (at 880µm). The radio polarisation is variable (typically within the ranges given before) on time scales of few hours. Interestingly, Macquart et al. (2006)) find the intrinsic angle of polarisation to be about 165°. This would be close ($\sim 25°$) to our result for the decay phase of the flare described in Sect. 2 (modulo 180°). Such an agreement could indicate that at least in some phases of activity NIR and radio observations are tracing emission from the same region around SgrA*.

Additional information has been found in the *circularly* polarised radio emission. This was originally reported by Bower et al. (1999b). Based on a re-analysis of elder VLA data, Bower (2003b) finds a constancy of the sign of the circular polarisation for about 20 years. This would – again – point towards a fixed B field orientation in the emission region.

- SgrA* flares show a quasi-periodic *sub-structure* on time scales of minutes. Examples for this are given in Fig. 3.7 (left column) where the lightcurves of five K_s band flares observed from 2003 to 2005 are presented. The first four (from top) panels show fluxes vs. time, the fifth panel shows the polarised flux of the flare described by Eckart et al. (2006b).

All flare lightcurves (especially panels 1, 2, 4) show characteristic structures: an overall profile (rise, maximum, decay) lasting about 1–2 hours is repeatedly modulated in cycles of 15–25 minutes.

In the right column of Fig. 3.7 Scargle periodograms (Scargle 1982) of the respective lightcurves are shown to visualise periodicities. The Scargle periodogram is defined as

$$P_X(\omega) = \frac{1}{2} \cdot \left[\frac{\left(\sum_j X_j \cos \omega t_j \right)^2}{\sum_j \cos^2 \omega t_j} + \frac{\left(\sum_j X_j \sin \omega t_j \right)^2}{\sum_j \sin^2 \omega t_j} \right] \qquad (3.2)$$

Here ω is the angular frequency, t_j the time of data point j, X_j is the value measured at time t_j, and P the power.

In order to illustrate the signatures presented in Fig. 3.7, Fig. 3.8 gives a simple model in form of an artificial lightcurve. Please note that this model is an illustration only, not a simulation or reconstruction of a flare. The artificial flare is composed of four additive components (in arbitary units): (a) a sine half wave with length 100 and amplitude 10 as overall profile, (b) a sine wave with amplitude 1.2 and period 18 as periodic modulation, (c) random Gaussian noise with $\sigma = 1$, and (d) a constant background of height 1. The sum of these four components forms the artificial flare. For each of the three non-constant components on the one hand and the final synthesised flare lightcurve on the other hand we computed the respective periodograms.

Comparing Figs. 3.7 and 3.8 allows to disentangle the features in the flare periodograms. These are: strong peaks at frequencies $\nu < 0.02$ 1/min due to the overall flare profiles; secondary maxima at $\nu \approx 0.04$–0.08 1/min due to the (quasi-)periodic substructure; and noise signals over the entire spectra.

Including this work, quasi-periodic signals in NIR flares have now been found in the range of 13–30 minutes. This sub-structure is generally quite weak – indeed

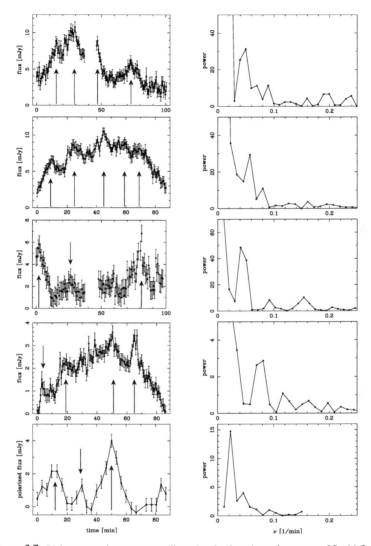

Figure 3.7: Lightcurves and power spectra illustrating the short-time sub-structure of SgrA* flares. Please note the different axes scales. The *left hand* panels show the observed light curves. All data were obtained with NACO at the VLT in K_s band. Local maxima in the overall flare shape are marked by arrows. Gaps in the light curves are due to intermediate sky observations. Times of observations are from top to bottom: June 15, 2003 (epoch 2003.45); June 16, 2003 (epoch 2003.45); July 6, 2004 (epoch 2004.51, first 100 min); July 8, 2004 (epoch 2004.52); July 29, 2005 (epoch 2005.57). The left bottom plot shows the polarised flux of the flare described by Eckart et al. (2006b). The *right hand* panels show the corresponding Scargle periodograms of the light curves given in the left column. In case of the polarised flux, the periodogram is dominated by the ∼40-min distance between the two strongest peaks. All graphs show a secondary maximum corresponding to periods about 13–25 min, with accuracies about 2 min.

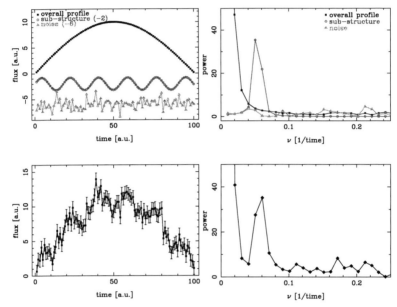

Figure 3.8: An artificial flare lightcurve as illustration of sub-structure signatures. All units are arbitary, but numerical values were selected thus that a comparison to Fig. 3.7 is straight forward. *Top left:* The additive components of the artificial flare: an overall profile (sine half-wave) with amplitude 10 and duration 100; a sine wave modulation with amplitude 1.2 and period 18; and Gaussian noise with $\sigma = 1$. *Top right:* Scargle periodograms for each of the three components shown in the top left panel. *Bottom left:* The synthetised flare lightcurve. This is the sum of the three components shown in the top left panel plus a constant background of height 1. To ease a comparison to real data, all data points are plotted with error bars with the length of the noise's σ. *Bottom right:* Scargle periodogram of the light curve shown in the bottom left panel.

the "double peak" of the flare described in sections 2 and 3 is the strongest case seen so far – and detected only in a part of all observed flares. This statement is true also for the X-ray activity of SgrA*, where quasi-periodic sub-structure with periods of about 5–22 min was reported for some flares.

3.5 Discussion

With all these elements at hand, we can start drawing fairly robust conclusions on the nature of the flares. However, at this point we do not conclude that we can derive reliable quantitative parameters of SgrA*. Many parameters would still highly depend on the model assumptions made by the author. Therefore such a quantitative statement would necessarily suffer from over-simplification. The important physical facts or hints are better obtained through qualitative discussion.

3.5.1 Nature of the flares

First of all, we know from imaging observations that flares occur always at the same location, consistent within a few mas (a few 100 Schwarzschild radii, R_S) with the gravitational centre of the nuclear starcluster and the radio source Sgr A* (Genzel et al. 2003b; Ghez et al. 2004; Eisenhauer et al. 2005a). But more importantly, the lightcurves of several NIR and X-ray flares observed so far show significant variations on the timescale of 15 min. This demonstrates that the region involved in these sub-structures is smaller than $\simeq 10 R_S$. Furthermore, the typical timescale in the lightcurves is consistent across these flares, ranging from 13 to 30 (± 2) min.

This timescale is within the range of the innermost stable circular orbit (ISCO; Bardeen et al. 1974) orbital periods allowed for Kerr black holes of 3–$4 \cdot 10^6 M_\odot$ and various spin parameters. Many authors (Genzel et al. 2003b; Yuan et al. 2004; Liu et al. 2004; Broderick & Loeb 2006; Paumard et al. *in prep.*) have studied the possibility that the flare emission may actually come from matter orbiting the BH close to the ISCO. The scatter in the observed periods is not a concern: in the context of this "orbiting blob scenario", it would simply indicate that the outbursts do not always occur exactly on the ISCO, but that a range of orbital radii is allowed. Orbits *inside* the ISCO are unstable. Since flares last for more than one orbital period (typically more than four), we can assume that flares occur *outside* this orbit. For this reason, the shortest period ever measured (13 ± 2 min; Fig. 3.7 and Table 3.4) sets a lower limit to the spin parameter a of the BH: using $M_{SgrA*} = 3.6 \pm 0.3 \cdot 10^6 M_\odot$ (Eisenhauer et al. 2005a), this leads to $a \geq 0.70 \pm 0.11$ (following Bardeen et al. 1974).

The presence of this quasi-periodic sub-structure imposes serious limits on alternative emission scenarii. *Bow shock fronts* caused by stars moving through the accretion disc material (Nayakshin et al. 2003) should not show such modulation.

In case of *jet emission* (Falcke & Markoff 2000; Markoff et al. 2001; Yuan et al. 2002) such modulations would be imprinted on the jet if the jet nozzle was located in the accretion disc, orbiting the black hole. Indeed, the jet model by Falcke & Markoff (2000) requires a nozzle with a radius of $\sim 4\,R_S$ and a height of $\sim 8\,R_S$. This extension would be small enough to allow for the observed short-time variability. On the other hand, Gillessen et al. (2006) analyse the cooling time scales of orbiting hot spots and find a limit on the extension of the emission region of $0.3\,R_S$. As this would be one order of magnitude smaller than the size of the model jet nozzle, the emission in the *early* phases of a flare probably cannot be explained by a "pure" jet emission model. This does not exclude the presence of a jet but makes it unlikely that a jet is responsible for a significant part of the observed NIR emission. In the *end* or *decay* phase of a flare, when a substantial shearing and broadening of the emission region is expected from the hotspot model, a growing contribution from resp. evolution into a jet is thinkable; we will pay attention to this again later.

Additionally, the observations also do not a priori exclude the possibility of spiral *density waves* propagating in the accretion disc. Those oscillations have been discussed in the context of stellar black hole systems in order to explain high-frequency (kHz range) QPOs (e.g. Kato 2001; Pétri 2006; and references therein). A rotating two-arm structure could double the orbital time scale and thus loose the constraints on the BH spin. But those structures are expected to have life times which are shorter than the time of a single orbital revolution (Schnittman et al. 2006); this does not agree with the observations of flares lasting for several orbital periods (up to hours).

Our description might be somewhat challenged by X-ray observations for which quasi-periodicities as short as 5 min have been claimed (Aschenbach et al. 2004; Aschenbach 2006). Using the dynamical picture described above, such a short period would require a spin of about $a = 0.99$. Indeed Aschenbach et al. (2004) claim the detection of several, resonant frequencies in the same flares. They interpret this as a signature of oscillations in the accretion disc, leading to a spin of $a = 0.996$ and a mass of $M_{SgrA*} = 3.3 \cdot 10^6 M_\odot$ (Aschenbach 2006). However, Bélanger et al. (2006) find one periodicity of 22 min and no additional signals in the same data sets. They developed a rigorous statistical method that excludes other (quasi-)periods being present to a statistically significant level in the data. This 22-min-period falls in the range of the periods observed in NIR flares. From now on, we implicitly assume that the flare emission comes from matter orbiting the BH.

3.5. Discussion

The fact that the flare emission is polarised nicely confirms the synchrotron radiation nature of the emitted light. This was already suspected from the overall spectral energy distribution, NIR and X-ray colour indices, and the occasional occurrence of simultaneous NIR and X-ray flares (Zylka & Mezger 1988; Zylka et al. 1992; Baganoff et al. 2001, 2003; Yuan et al. 2004; Liu et al. 2004; Eckart et al. 2006a). In this context, the polarisation parameter curves (Figs. 3.3, 3.5) convey information about the geometry of the magnetic field. The remarkable permanence of the polarisation parameters, in particular polarisation angle, across three NIR flares observed over 2 years (Eckart et al. 2006b; this work) indicates that the magnetic field geometry as well as the orbital plane remained the same for all three events. This shows that the flaring material has enough time to settle in the BH's equatorial plane before the occurrence of the flare, and speaks in favour of a somewhat permanent accretion disc experiencing energetic events rather than temporary structures building up randomly for each flare.

3.5.2 Geometry of the system

When comparing our data to the models by Broderick & Loeb (2006), the non-detection of variations in the polarisation *angle* during the peak phase strongly suggests that the accretion disc is seen (within few degrees) edge-on. The non-detection of variations in the polarisation *fraction* also speaks for an edge-on view of the disc. This parameter is not exactly constant, but it would show only a short dip ($\sim$2 min) that our time sampling would not allow detecting. Additionally, the constancy of the polarisation fraction (about 15% during the peak phase) speaks against the picture of (a) a dominating, slow component of the lightcurves (e.g. emission from the disc itself) on top of which (b) the periodic signal due to a second component is seen. Such a second component would be unlikely to be subject to the same magnetic field as the bright spot itself. On the contrary, the mechanism proposed by Paumard et al. *in prep.* by which this slow component is due to shearing of the hotspot, evolving into a ring, fits this observational result well.

There are basically two possible geometries for the magnetic field in the orbiting spot scenario: poloidal (perpendicular to the orbital plane) and toroidal (tangential to the orbit). The poloidal field is more natural in the absence of matter or outside of the disc. On the other end, the field inside the disc is most likely frozen and dragged by the matter. Due to shear, this naturally leads to a toroidal field (De Villiers et al. 2003; Broderick & Loeb 2006). A transition region above the disc and at its inner edge must exist, in which the magnetic field is somewhat disorganised. This explains the fairly low observed polarisation fraction ($\sim$15%; in a perfectly organised field, the polarisation fraction of synchrotron emission is of order 75%, Pacholczyk 1970). The question remains which component of the field is dominant; we will come back to this later.

The decay part of the flare reported here (sections 2, 3) shows a dramatic change in both polarisation fraction (from 15% to 40%) and polarisation angle (from 80° to 10°). It follows that the magnetic field seen by the electrons also changes dramatically. It becomes much more organised, leading to an increased polarisation fraction, and rotates by $\simeq 70°$. There are two options to explain this change: either the field changes where the electrons are, or the electrons move to region with a different field geometry. We will explore both possibilities below.

Let us first assume the flaring material remains on the orbital plane: in this case, a change in the magnetic field could be due to the fact that the accretion disc vanishes, letting the magnetic field relax into its matter-less state, which is poloidal. This means that the field was mostly toroidal during the peak phase. The same conclusion is reached if the matter leaves the disc from its inner edge, falling onto the BH.

The other possibility is that the material moves out of the accretion disc. Since the flares are magnetically driven, it seems natural to assume that this matter would follow the field lines, perhaps into a jet. Here again, the final magnetic field is likely poloidal, hence the initial field is toroidal.

In these two schemes, the field is toroidal during the peak phase and poloidal during the decay phase. We now assume that this is the case. A toroidal field in the peak phase is yet another hint that the material has settled into a disc and been able to drag the field before the occurrence of the flare.

The orientation of the magnetic field with respect to the Galactic plane, which is located at $+27°$, contains additional information. Indeed the peak phase polarisation angle is roughly perpendicular to the Galactic plane (to within ~ 30 degrees), whereas the decay phase polarisation angle is mostly in the Galactic plane (to within ~ 10 degrees). We therefore see here an indication that the accretion disc of Sgr A* lies essentially in the plane of the Galaxy, and that its spin axis is essentially aligned with that of the Galaxy. But as long as there are no stricter constraints on the polarised NIR emission from SgrA*, it is possible that future observations revise this picture.

3.5.3 Proposed model

Given together, we state that our data support the following model: SgrA* is a fairly rapidly (perhaps maximally) rotating BH. Its spin axis is essentially aligned with that of the Galaxy. It is surrounded by a somewhat permanent accretion disc, with an inner edge close to the ISCO, in which the magnetic field is toroidal. Outside of this disc, the field is poloidal. Occasionally, shear will bend the magnetic field so much that a magnetic reconnection is warranted. This is most likely to occur near the inner edge of the disc, where shear is most effective. The magnetic reconnection heats a fraction of the electrons to a hot temperature ($\simeq 10^{12}$ K). The region affected is localised, smaller than the constraint imposed by cooling-time arguments in Gillessen et al. (2006): $R < 0.3 R_S$. These electrons swirl in the toroidal magnetic field and emit synchrotron emission. The emitting region orbits the BH, giving raise to the periodic signal we observe. Shear as well as magnetic forces make the region extend along the orbit. Since it spans only a small range in distance from the BH, the shear is not extremely fast and allows the periodic signal to be discernable for several periods. Nevertheless, within a few orbital periods, the entire ISCO glows in synchrotron emission, and this emission is responsible for the dominating, slow part of the lightcurves (Paumard et al. *in prep.*). After some time, the magnetic reconnection is over, removing the heating mechanism from the picture. The electron population cools down, and at the same time extend outside of the disc, perhaps into a jet. The dominating field then becomes poloidal.

3.6 Summary

On May 31st, 2006, we observed a K_s band flare from SgrA* which shows

- a high level of total flux, up to $\sim$16 mJy;
- strong, variable polarisation, with $p = 15...40\%$;
- a polarisation angle between $\sim 80°$ (during the peak phase) and $\sim 10°$ (in the decay phase), swinging within about 15 minutes;
- repeated sub-structure (double peak in total and polarised flux) on a time scale of 15 min.

Using this as well as information gathered during the last years from radio, NIR, and X-ray observations, we see strong indication that the flare emission in SgrA* is the synchrotron emission from material orbiting the BH. We also find indication that some of this material eventually makes it into a jet, reconciling the "orbiting spot scenario" tenants with the jet hypothesis literature (Markoff et al. 2001; Yuan et al. 2002).

Finally, we might have observed the first pieces of evidence that a SMBH spin axis is aligned with that of its host galaxy.

Acknowledgements: Special thanks to N. Ageorges, ESO, for helpful discussions on NACO. We are grateful to the ESO instrument scientists and engineers who made possible this successful work. F.M. acknowledges support from the Alexander von Humboldt Foundation. We also would like to thank the anonymous reviewer whose comments helped to improve the quality of this article.

Chapter 4

Motion of the stellar cusp

Original publication: M.J. Reid, K.M. Menten, S. Trippe, T. Ott & R. Genzel 2007, *The Position of Sagittarius A*: III. Motion of the Stellar Cusp*, ApJ, 659, 378

Abstract: In the first two papers of this series (i.e. Reid et al. 1999, 2003), we determined the position of SgrA* on infrared images, by aligning the positions of red giant stars with positions measured at radio wavelengths for their circumstellar SiO masers. In this paper, we report detections of 5 new stellar SiO masers within 50" (2 pc) of SgrA* and new and/or improved positions and proper motions of 15 stellar SiO masers. The current accuracies are ≈ 1 mas in position and ≈ 0.3 mas/yr in proper motion. We find that the proper motion of the central stellar cluster with respect to SgrA* is less than 45 km/s. One star, IRS 9, has a three-dimensional speed of ≈ 370 km/s at a projected distance of 0.33 pc from SgrA*. If IRS 9 is bound to the inner parsec, this requires an enclosed mass that exceeds current estimates of the sum of the mass of SgrA* and luminous stars in the stellar cusp by $\approx 0.8 \times 10^6$ $M_\odot$. Possible explanations include i) that IRS 9 is not bound to the central parsec and has "fallen" from a radius greater than 9 pc, ii) that a cluster of dark stellar remnants accounts for some of the excess mass, and/or iii) that R_0 is considerably greater than 8 kpc.

4.1 Introduction

Sagittarius A* (SgrA*), the compact radio source at the center of the Galaxy (Balick & Brown 1974), is almost surely a super-massive black hole. Infrared-bright stars orbit about the radio position of SgrA* and require $\approx 3.9 \times 10^6$ $M_\odot$ within a radius of ≈ 50 AU (Eisenhauer et al. 2005a; Ghez et al. 2005a). While orbiting stars move at thousands of km/s, SgrA* is essentially stationary at the dynamical center of the Galaxy, moving < 2 km/s out of the Galactic plane (Backer & Sramek 1999; Reid et al. 1999; Reid & Brunthaler 2004). This indicates that the radiative source SgrA*, which is less than 1 AU in size (Rogers et al. 2004; Krichbaum et al. 1998; Doeleman et al. 2001; Bower et al. 2004; Shen et al. 2005), contains a significant fraction ($> 10\%$) of the gravitational mass (Reid & Brunthaler 2004).

The Galactic Center stellar cluster contains red giant stars that are both strong radio sources (from circumstellar SiO maser emission) and bright infrared sources. Because these stars are visible at both radio and infrared wavelengths, one can use their radio positions, measured with respect to SgrA*, to determine accurate scale, rotation, and distortion corrections for an infrared image. This allows the highly accurate radio reference frame to be transfered to the infrared images, improving the quality of the infrared astrometry and precisely locating the position of SgrA*. In Menten et al. (1997) and Reid et al. (2003), hereafter Papers I & II, we developed this technique and determined the position of SgrA* on diffraction-limited 2.2 μm wavelength images of the Galactic Center with an accuracy of ≈ 15 mas. Locating SgrA* on infrared images has been important for determining its emission in the presence of confusing stellar sources and verifying that foci of stellar orbits coincide with the position of SgrA*. This links the *radiative* (compact radio) source with the *gravitational* source.

We present new VLA observations of stellar SiO masers in the central cluster, updating their positions and, by more than doubling the observing time-span, significantly improving their proper motion determinations. We place the proper motions of infrared stars in the central cluster in a reference frame tied to SgrA*. In sect. 2 we describe the radio measurements of the positions and proper motions of 15 SiO maser stars, and in sect. 3 we present the latest infrared positions and proper motions of those stars within 20 arcsec of SgrA*. We use these data to transfer the infrared proper motions to a reference frame tied to SgrA* in sect. 4. Finally, in sect. 5, we use the three-dimensional speeds and projected offsets of stars from SgrA* to constrain the combined mass of SgrA* and the central (luminous and dark) stellar cluster.

4.2 Radio Observations

Over the period 1995 to 2006 we have searched for and mapped SiO masers associated with late-type stars that are projected near SgrA*. We used the NRAO[1] VLBA and VLA to measure accurately the relative positions of SiO maser stars and SgrA*.

Red giant and super-giant stars of late-M spectral class often exhibit SiO masers in their extended atmospheres. These masers are strongly variable over time scales of

---

[1]NRAO is a facility of the National Science Foundation operated under cooperative agreement by Associated Universities, Inc.

4.2. Radio Observations

~ 1 y. SiO maser emission emanates from a typical radius of ~ 4 AU (e.g. Diamond & Kemball 2003), which corresponds to ~ 0.5 mas at the assumed 8.0 kpc distance of the Galactic Center (Reid 1993). As our measurements were made over a time-span much longer than the stellar cycle, we cannot track individual maser features, and we accept an intrinsic stellar position uncertainty of about ±0.5 mas owing to possible variations across the maser shell. For a late-type super-giant, the stellar radius is considerably larger than for a red giant of similar spectral class, and its SiO masers are found at radii roughly an order of magnitude larger. Variation of the SiO masers in a super-giant can considerably degrade inferred stellar position and proper motion measurements (see discussion of IRS 7 in sect. 2.1).

4.2.1 VLA Observations

Our VLA observations were conducted in the A-configuration in 1995 June (reported in Paper I), 1998 May and 2000 October/November (reported in Paper II) and 2006 March (reported here). Near the Galactic Center, SiO masers are likely detectable over a very wide range of velocities, probably exceeding 700 km/s. However, wide-band observations at the VLA are currently severely limited by the correlator. In order to obtain adequate spectral resolution and sensitivity, we chose to limit our velocity coverage and observe in seven 6.25 MHz bands (each covering $\approx$ 40 km/s excluding band edges). We observed in both right and left circular polarization for each band and obtained 64 spectral channels per band, resulting in uniformly weighted spectral channel spacings of about 98 kHz or 0.67 km/s.

Our 2006 observations were conducted on March 5, 18, & 19. The VLA had 24 antennas in operation, and the synthesized beam toward SgrA* was about 86×33 mas elongated in the north-south direction. Except for occasional calibration sources, we pointed on the position of SgrA*, allowing detection of masers within the primary beam of a VLA antenna ($\approx 30''$ HWHM at 43 GHz). We observed by cycling among bands centered at LSR velocities of $-346, -111, -73, -39, -1, +40$ and $+75$ km/s; the latter six bands covered the LSR velocity range -131 to $+95$ km/s with only two small gaps. This setup allowed for deeper integrations for the previously known, or suspected, SiO maser stars than would be possible for a "wide-open" search.

Initial calibration of the VLA data was done in a standard manner recommended by the Astronomical Image Processing System (AIPS) documentation. The flux density scale was based on observations of 3C 286, assuming 1.49 Jy for interferometer baselines shorter than 300 kλ. Amplitude and bandpass calibration was accomplished with observations of NRAO 530, which had a flux density of 2.8 Jy during the 2006 observations. The visibility data were then self-calibrated (amplitude and phase) on

SgrA* for each individual 10-sec integration. This removed essentially all interferometer phase fluctuations, owing to propagation through the Earth's atmosphere, and placed the phase center at the position of SgrA*.

SgrA* is known to vary occassionally in flux density by $\approx$ 10% on hourly time scales (Yusef-Zadeh et al. 2006b). Were this to happen during our observations, self-calibrating on SgrA* would introduce false gain (amplitude) variations of a similar magnitude and degrade somewhat the dynamic range of the SiO maser images. However, this is unlikely to shift the measured position of the masers significantly, as the positions are determined almost entirely by the phase data.

We searched for new maser stars by making very large images, covering about ±50 arcsec, or most of the primary beam of an individual VLA antenna at 43 GHz, about SgrA*. This was done by limiting the range of uv-data used and resulted in maps with a nearly circular beam of 0.15 arcsec. Typical rms noise levels in these images were near 9 mJy, allowing 6σ detections of 54 mJy. Five new SiO maser stars were discovered: SiO-14, SiO-15, SiO-16, SiO-17, and IRS 19NW.

Once the approximate location of a maser was known, either from previous observations or from the large images, we mapped each band with up to five small sub-images centered on known or suspected masers with emission in that band. These synthesized maps typically had single spectral-channel noise levels of about 5 mJy. We always included a sub-image for SgrA* at the phase center of the interferometric data. By simultaneously imaging the stellar SiO masers and the continuum emission from SgrA*, the strong continuum emission from SgrA* did not degrade the detections of relatively weak SiO masers far from SgrA*. A composite spectrum from all seven observing bands for the 1998, 2000, and 2006 observations is shown in Fig. 4.1.

As described in Paper II, we obtained a single position for each star at each observing epoch by 1) fitting a 2-dimensional Gaussian brightness distribution to each spectral channel with detectable SiO emission, 2) averaging, using variance weighting, to obtain a best stellar position and estimated uncertainty (if the reduced χ^2 was greater than unity, we increased the position uncertainties accordingly), and 3) correcting for differential aberration, an effect of < 1 mas for stars < 15 arcsec of SgrA*.

We list the positions of the SiO maser stars, relative to SgrA*, in Table 4.2.1. We include the results from the 1995 VLA observations reported in Paper I, the 1998 and 2000 VLA observations reported in Paper II, and the 2006 results reported in this paper. Since stellar SiO masers are variable in strength over the period of the stellar pulsation and the sensitivity of each epoch's data differed somewhat, only the stronger sources are detected at all epochs.

One of the stars, IRS 7, is a super-giant. As discussed in Paper II, it has SiO maser

4.2. Radio Observations

Star	V_{LSR} (km/s)	$\Delta\Theta_x$ (arcsec)	$\Delta\Theta_y$ (arcsec)	Epoch (y)	Telescope
IRS 9	−342	5.6515 ± 0.0048	−6.3589 ± 0.0060	1998.39	VLA
		5.6501 ± 0.0007	−6.3509 ± 0.0013	1998.41	VLA
		5.6589 ± 0.0011	−6.3454 ± 0.0017	2000.85	VLA
		5.6742 ± 0.0004	−6.3347 ± 0.0009	2006.20	VLA
IRS 7	−114	0.0403 ± 0.0080	5.5829 ± 0.0130	1995.49	VLA
		0.0387 ± 0.0023	5.5676 ± 0.0049	1998.39	VLA
		0.0378 ± 0.0043	5.5495 ± 0.0014	1998.41	VLA
		0.0342 ± 0.0016	5.5414 ± 0.0030	2000.85	VLA
		0.0326 ± 0.0007	5.5238 ± 0.0013	2006.20	VLA
SiO-14	−112	−7.6648 ± 0.0012	−28.4528 ± 0.0020	1998.41	VLA
		−7.6596 ± 0.0005	−28.4530 ± 0.0008	2000.85	VLA
		−7.6485 ± 0.0005	−28.4586 ± 0.0009	2006.20	VLA
IRS 12N	−63	−3.2519 ± 0.0005	−6.8814 ± 0.0005	1996.41	VLBA
		−3.2541 ± 0.0005	−6.8877 ± 0.0006	1998.39	VLA
		−3.2543 ± 0.0010	−6.8860 ± 0.0011	1998.41	VLA
		−3.2554 ± 0.0009	−6.8936 ± 0.0012	2000.85	VLA
		−3.2626 ± 0.0009	−6.9073 ± 0.0019	2006.20	VLA
IRS 28	−55	10.4702 ± 0.0030	−5.7884 ± 0.0050	1998.41	VLA
		10.4693 ± 0.0010	−5.7956 ± 0.0024	2000.85	VLA
		10.4809 ± 0.0007	−5.8254 ± 0.0010	2006.20	VLA
SiO-15	−36	−12.4253 ± 0.0023	−11.0794 ± 0.0054	2000.85	VLA
		−12.4385 ± 0.0012	−11.0753 ± 0.0015	2006.20	VLA
IRS 10EE	−27	7.6821 ± 0.0030	4.2125 ± 0.0050	1995.49	VLA
		7.6837 ± 0.0005	4.2194 ± 0.0005	1996.41	VLBA
		7.6841 ± 0.0005	4.2146 ± 0.0009	1998.39	VLA
		7.6837 ± 0.0005	4.2157 ± 0.0005	1998.41	VLA
		7.6845 ± 0.0005	4.2099 ± 0.0005	2000.85	VLA
		7.6840 ± 0.0005	4.1990 ± 0.0005	2006.20	VLA
IRS 15NE	−12	1.2256 ± 0.0140	11.3108 ± 0.0210	1995.49	VLA
		1.2302 ± 0.0005	11.3315 ± 0.0005	1996.41	VLBA
		1.2249 ± 0.0017	11.3193 ± 0.0019	1998.39	VLA
		1.2270 ± 0.0005	11.3201 ± 0.0006	1998.41	VLA
		1.2228 ± 0.0011	11.3024 ± 0.0025	2000.85	VLA
		1.2112 ± 0.0005	11.2761 ± 0.0010	2006.20	VLA
SiO-16	+8	−26.4237 ± 0.0020	−34.4520 ± 0.0027	1998.41	VLA
		−26.4216 ± 0.0006	−34.4412 ± 0.0011	2000.85	VLA
		−26.4191 ± 0.0006	−34.4553 ± 0.0013	2006.20	VLA
SiO-6	+52	35.1982 ± 0.0090	30.6567 ± 0.0140	1995.49	VLA
		35.2207 ± 0.0029	30.6537 ± 0.0050	1998.39	VLA
		35.2323 ± 0.0006	30.6593 ± 0.0010	2000.85	VLA
		35.2451 ± 0.0010	30.6702 ± 0.0028	2006.20	VLA
SiO-17	+53	8.0427 ± 0.0005	−27.7034 ± 0.0014	1998.41	VLA
		8.0624 ± 0.0005	−27.6852 ± 0.0009	2006.20	VLA
SiO-11	+70	1.7121 ± 0.0040	40.2614 ± 0.0060	1995.49	VLA
		1.7379 ± 0.0023	40.2681 ± 0.0032	1998.39	VLA
		1.7401 ± 0.0005	40.2794 ± 0.0011	2000.85	VLA
		1.7462 ± 0.0005	40.2914 ± 0.0006	2006.20	VLA
IRS 17	+73	13.1501 ± 0.0026	5.5651 ± 0.0025	2000.85	VLA
		13.1415 ± 0.0013	5.5611 ± 0.0021	2006.20	VLA
SiO-12	+82	−18.8645 ± 0.0190	42.4905 ± 0.0290	1995.49	VLA
		−18.8235 ± 0.0028	42.4686 ± 0.0032	2000.85	VLA
		−18.8212 ± 0.0017	42.4459 ± 0.0033	2006.20	VLA
IRS 19NW	+84	14.5518 ± 0.0011	−18.4619 ± 0.0012	1998.41	VLA
		14.5532 ± 0.0015	−18.4683 ± 0.0031	2000.85	VLA
		14.5607 ± 0.0005	−18.4656 ± 0.0009	2006.20	VLA

Table 4.1: SiO maser astrometry. VLBA positions are reported at a single reference epoch. VLA data have been corrected for differential aberration. $\Delta\Theta_x$ and $\Delta\Theta_y$ are angular offsets relative to SgrA* toward the east and north, respectively, in the J2000 coordinate system.

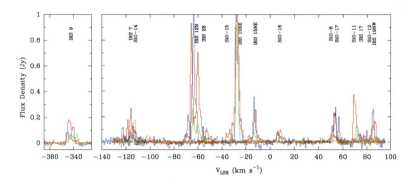

Figure 4.1: Composite spectrum of stellar SiO masers detected with the VLA in 1998 (*blue*), 2000 (*green*) and 2006 (*red*) observations. Stars are identified at the top of the spectrum at their approximate stellar velocities.

features spread over ≈ 20 km/s and is highly variable. We would expect its stronger SiO maser peaks could be spread over a region of at least 10 mas. Thus, the positions determined from the brightest SiO maser feature(s) in IRS 7 may not indicate the stellar position to better than about 5 mas, and we have increased the position uncertainties for IRS 7 in Table 4.2.1 to allow for this possibility.

We constructed spectra at the pixel of peak brightness for SiO masers detected in the 1998, 2000, and 2006 VLA observations. These spectra are displayed in Figs. 4.2 & 4.3. Most of these SiO spectra are as expected for Mira variables located at the distance of the the Galactic Center: they show flux densities ≤ 1 Jy covering a velocity range of 5 to 10 km/s and strong variability over timescales of years. Fig. 4.4 shows the Galactic Center region, with the positions and proper motions of the nine SiO maser stars that are projected within the inner 21" displayed.

4.2.2 SiO Maser Proper Motions

We determined stellar proper motions by fitting a variance-weighted straight line to the positions as a function of time from all of the available data compiled in Table 4.2.1. These proper motion fits are given in Table 4.2.1 and displayed graphically in Figs. 4.2 & 4.3. The reference epoch for the proper motion solution was chosen as the variance-weighted mean epoch for each star, in order to obtain uncorrelated position and motion estimates. Since the estimated uncertainties for individual east-west and north-south positions were neither identical, nor exactly linearly related, we chose a single, average reference epoch for each star (instead of a separate reference epoch for the east-west and north-south directions), which resulted in slight parameter correlations.

4.2. Radio Observations

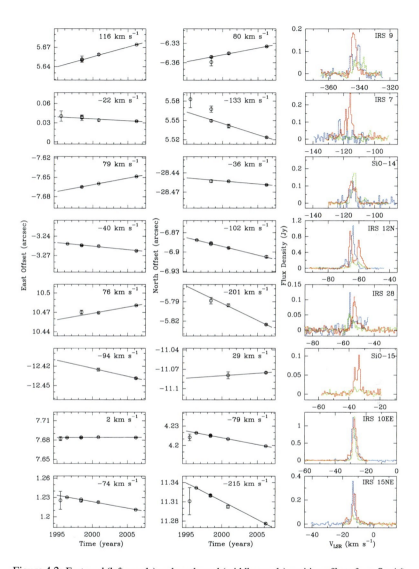

Figure 4.2: Eastward (left panels) and northward (middle panels) position offsets from SgrA* versus time for the eight SiO maser stars with negative LSR velocities. Solid lines are variance-weighted best-fit proper motions. The linear speed is indicated in each frame, assuming a distance of 8.0 kpc. Spectra from 1998 (*blue*), 2000 (*green*) and 2006 (*red*) observations are also shown (right panels). Star names are indicated in the upper right corner of the right panels.

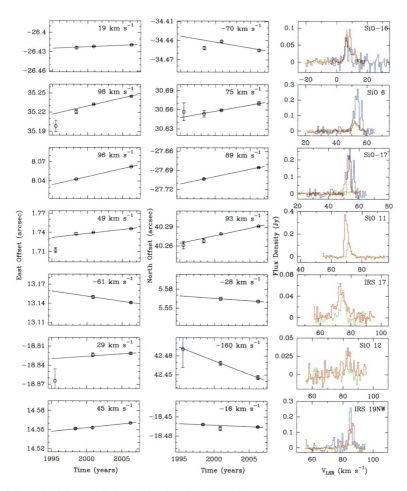

Figure 4.3: Eastward (left panels) and northward (middle panels) position offsets from SgrA* versus time for the seven SiO maser stars with positive LSR velocities. Solid lines are variance-weighted best-fit proper motions. The linear speed is indicated in each frame, assuming a distance of 8.0 kpc. Spectra from 1998 (*blue*), 2000 (*green*) and 2006 (*red*) observations are also shown (right panels). Star names are indicated in the upper right corner of the right panels.

4.2. Radio Observations

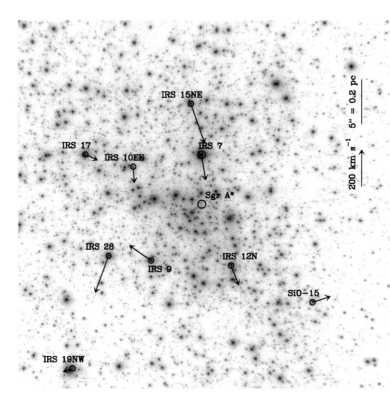

Figure 4.4: Infrared (K-band) image of the central ±20" of the Galactic Center taken in 2005, with east to the left and north to the top. SiO maser stars within this region are circled and their proper motions relative to SgrA* are indicated with arrows. The vertical bar and arrow at the right of the image indicate the linear and motion scales for $R_0 = 8.0$ kpc. The location of SgrA* is indicated by the circle at the center of the image.

Star	V_{LSR} (km/s)	$\Delta\Theta_x$ (arcsec)	$\Delta\Theta_y$ (arcsec)	μ_x (mas/yr)	μ_y (mas/yr)	Epoch (y)	Number Epochs
IRS 9	−342	+5.6655 ± 0.0003	−6.3407 ± 0.0007	+3.06 ± 0.10	+2.11 ± 0.19	2003.34	3
IRS 7	−114	+0.0336 ± 0.0050	+5.5401 ± 0.0050	−0.58 ± 0.50	−3.52 ± 0.54	2004.37	4
SiO-14	−112	−7.6554 ± 0.0003	−28.4553 ± 0.0006	+2.08 ± 0.12	−0.94 ± 0.20	2002.89	3
IRS 12N	−63	−3.2537 ± 0.0003	−6.8864 ± 0.0003	−1.06 ± 0.10	−2.70 ± 0.17	1998.17	4
IRS 28	−55	+10.4784 ± 0.0011	−5.8190 ± 0.0010	+2.00 ± 0.38	−5.29 ± 0.42	2005.00	3
SiO-15	−36	−12.4372 ± 0.0011	−11.0757 ± 0.0015	−2.47 ± 0.98	+0.77 ± 2.10	2005.68	2
IRS 10EE	−27	+7.6840 ± 0.0003	+4.2114 ± 0.0003	+0.04 ± 0.07	−2.09 ± 0.07	2000.24	5
IRS 15NE	−12	+1.2257 ± 0.0003	+11.3171 ± 0.0004	−1.96 ± 0.07	−5.68 ± 0.12	1998.92	5
SiO-16	+8	−26.4207 ± 0.0004	−34.4478 ± 0.0043	+0.49 ± 0.15	−1.84 ± 1.52	2002.82	3
SiO-6	+52	+35.2333 ± 0.0011	+30.6605 ± 0.0009	+2.58 ± 0.43	+1.99 ± 0.52	2001.43	4
SiO-17	+53	+8.0560 ± 0.0004	−27.6911 ± 0.0008	+2.53 ± 0.18	+2.34 ± 0.42	2003.68	2
SiO-11	+70	+1.7441 ± 0.0014	+40.2871 ± 0.0007	+1.30 ± 0.46	+2.45 ± 0.25	2004.38	4
IRS 17	+73	+13.1442 ± 0.0012	+5.5624 ± 0.0016	−1.61 ± 1.08	−0.75 ± 1.22	2004.49	2
SiO-12	+82	−18.8227 ± 0.0030	+42.4559 ± 0.0023	+0.77 ± 1.14	−4.24 ± 0.84	2003.83	3
IRS 19NW	+84	+14.5578 ± 0.0005	−18.4647 ± 0.0012	+1.19 ± 0.14	−0.43 ± 0.31	2003.79	3

Table 4.2: SiO maser proper motions. For sources with VLBA detections, only a single position was used when fitting for proper motions. VLA data have been corrected for differential aberration. $\Delta\Theta_x$ and $\Delta\Theta_y$ are angular offsets at the listed epoch, and μ_x and μ_y are proper motions, relative to SgrA* toward the east and north, respectively, in the J2000 coordinate system. IRS 7 was shifted by +0.010" northward to "center the star" between two maser positions; its position and proper motion uncertainties were increased to 0.005" and 0.5 mas/yr. Formal proper motion uncertainties were doubled for the stars with only 2-epoch motions.

Assuming a distance of 8.0 kpc to the Galactic Center (Reid 1993), we convert proper motions to linear velocities and, with the radial velocities, determine the full 3-dimensional speed, V_{total}, of each star with respect to SgrA*. These speeds are given in Table 4.2.2. While the speeds on the plane of the sky are directly referenced to SgrA*, the speeds along our line-of-sight are in the LSR reference frame. Thus, our values of V_{total} assume that SgrA* has a near-zero line-of-sight speed with respect to the LSR. As no spectral lines have been detected from SgrA*, there is no direct observational evidence supporting this assumption. However, the lack of a detectable proper motion of SgrA* suggests that it anchors the dynamical center of the Galaxy (Reid & Brunthaler 2004) and should be nearly at rest in the LSR frame.

4.3 Infrared Observations

Near-IR observations were obtained on the 8.2-m UT4 (Yepun) of the ESO-VLT on Cerro Paranal, Chile, using the detector system NAOS/CONICA (hereafter NACO) consisting of the adaptive optics system NAOS (Rousset et al. 2003) and the 1024 × 1024-pixel near-IR camera CONICA (Hartung et al. 2003b). We obtained 8 data sets in H and K bands with a scale of 27 mas/pixel (large scale) covering 5 epochs (May 2002, May 2003, June 2004, May 2005, April 2006).

Each image covers a field of view (FOV) of 28 × 28 arcsec. During each observa-

4.3. Infrared Observations

Star	$V_{\rm LSR}$ (km/s)	V_x (km/s)	V_y (km/s)	V_{total} (km/s)	R_{proj} (pc)	M_{encl} ($10^6\,M_\odot$)
IRS 9 ...	-342 ± 3	116 ± 4	80 ± 7	370 ± 3	0.33	>5.1
IRS 7 ...	-114 ± 3	-22 ± 19	-133 ± 20	177 ± 16	0.21	>0.5
SiO-14 ...	-112 ± 3	79 ± 5	-36 ± 8	142 ± 4	1.14	>2.4
IRS 12N ...	-63 ± 3	-40 ± 4	-102 ± 6	127 ± 6	0.30	>0.5
IRS 28 ...	-55 ± 3	76 ± 14	-201 ± 16	221 ± 15	0.46	>2.0
SiO-15 ...	-36 ± 3	-94 ± 37	29 ± 80	105 ± 40	0.65	>0.0
IRS 10EE ...	-27 ± 3	-5 ± 3	-82 ± 3	87 ± 3	0.34	>0.3
IRS 15NE ...	-12 ± 3	-74 ± 5	-215 ± 5	228 ± 4	0.44	>2.5
SiO-16 ...	8 ± 3	19 ± 6	-70 ± 58	73 ± 55	1.68	>0.0
SiO-6 ...	52 ± 3	98 ± 16	75 ± 20	134 ± 16	1.81	>2.2
SiO-17 ...	53 ± 3	96 ± 3	89 ± 8	141 ± 6	1.12	>2.2
SiO-11 ...	70 ± 3	49 ± 17	93 ± 9	126 ± 10	1.56	>2.1
IRS 17 ...	73 ± 3	-61 ± 41	-28 ± 46	99 ± 29	0.55	>0.1
SiO-12 ...	82 ± 3	29 ± 43	-160 ± 32	183 ± 29	1.80	>3.3
IRS 19NW ...	84 ± 3	45 ± 5	-16 ± 12	97 ± 4	0.91	>0.8

Table 4.3: 3-dimensional stellar motions and enclosed mass limits. V_x and V_y are proper motions speeds toward the East and North, respectively. $V_{total} = \sqrt{V_{\rm LSR}^2 + V_x^2 + V_y^2}$ is the total speed of the stars relative to SgrA*. Proper motion speeds, projected distances, total speeds and enclosed mass limits assume a distance to the Galactic Center of 8.0 kpc.

tion the camera pointing was shifted so that the total FOV of a complete data set was between 35×35 and 40×40 arcsec, centered on SgrA*. In all cases the spatial resolution was (nearly) diffraction limited, leading to a typical FWHM of ~ 60 mas. Typical limiting magnitudes were 18th mag at K band and 20th mag at H band. All images were sky-subtracted, bad-pixel removed and flat-field corrected. In order to obtain the best signal-to-noise ratios and maximum FOV coverages in single maps, we combined all good-quality images belonging to the same data set to mosaics after correcting for instrumental geometric distortion. Details of the distortion correction will be given in Trippe et al. (in prep.).

In order to establish an astrometric near-IR reference frame, we selected one high-quality mosaic (May 2005), extracted image positions for all detected stars and transformed them into astrometric coordinates using the positions of all 9 maser stars in the FOV interpolated to the epoch of the image. To compare positions among images of different epochs, we chose an ensemble of ≈ 600 "well-behaved" stars (i.e. stars that are well-separated from neighbors and are bright but not saturated) in the "master mosaic" and computed in each image all stellar positions relative to this ensemble. Due to varying FOVs, the number of stars usable for a given mosaic varied between about 400 to 600.

In each image, positions were extracted by fitting stars with 2-dimensional elliptical Gaussian brightness distributions. Although over the entire FOV significant departures from isoplanicity occur, this effect elongates the stellar PSFs symmetrically and does not affect significantly the centroids of Gaussian-fitted positions. Proper motions were computed by fitting the positions versus times with straight lines. To assure statistically "clean" errors for the proper motions, outlier rejection and error rescaling were applied where possible. This led to typical position accuracies of ≈ 1 mas and typical proper motion accuracies of ≈ 0.3 mas/yr. Unfortunately, this accuracy was not achieved for all maser stars, as some are very bright stars and are affected by detector non-linearity/saturation in some images; also the star most distant from SgrA* (IRS19NW) was observed only in the last two epochs. Thus, some of the maser stars have errors larger than typical.

4.4 Radio & Infrared Frame Alignments

The proper motions of stars in the Galactic Center cluster, measured from infrared images, are *relative* motions only. One can add an arbitrary constant vector to all of the stellar proper motions without violating observational constraints. Until now, the "zero points" of the motions have been determined by assuming isotropy and removing the average motion of the entire sample. Since the radio proper motions are inherently in a reference frame tied directly to SgrA*, one can use any one of the SiO stars, or the mean motion of a group of them, to place the infrared proper motions in SgrA*'s frame.

We have measured radio positions and proper motions, relative to SgrA*, for the nine bright SiO maser stars that appear on the NACO images (ie, within ≈ 20 arcsec of SgrA*). The position and proper motion accuracies typically are ~ 1 mas and ~ 0.3 mas/yr, respectively. This allows us to align the radio and infrared frames, both in position and in proper motion. Thus, not only can the location of SgrA* can be accurately determined on infrared images, but also stellar proper motions from infrared data can be referenced directly to SgrA*, without assumptions of isotropy or homogeneity of the stellar motions.

The radio and infrared proper motions measured for the nine stars are listed in Table 4.4. The nine stars have weighted mean differences (and standard errors of the means) of $+0.66 \pm 0.21$ mas/yr toward the east, and -0.45 ± 0.28 mas/yr toward the north. These results are qualitatively similar to those published in Paper II. Quantitatively, the differences between the IR motions of Paper II and this paper for some stars (notably IRS 12N) are greater than expected based on the quoted uncertainties. Since the IR motions in Paper II were based on 2 epochs only, which did not allow for an internal check on the formal motion uncertainties, we believe those uncertainties were somewhat optimistic.

Currently only one star, IRS 7, has a significant discrepancy between the radio and infrared motions in the both coordinates. This is the only super-giant star in the sample

4.5. Enclosed Mass versus Radius from SgrA*

Star	μ_x^{Radio} (mas/yr)	μ_y^{Radio} (mas/yr)	μ_x^{IR} (mas/yr)	μ_y^{IR} (mas/yr)	μ_x^{Dif} (mas/yr)	μ_y^{Dif} (mas/yr)
IRS9 ...	3.06 ± 0.10	2.11 ± 0.19	4.44 ± 0.55	1.10 ± 0.58	1.38 ± 0.56	−1.01 ± 0.61
IRS7 ...	−0.58 ± 0.50	−3.52 ± 0.54	1.86 ± 1.18	−7.33 ± 0.72	2.44 ± 1.28	−3.81 ± 0.90
IRS12N ...	−1.06 ± 0.10	−2.70 ± 0.17	−0.77 ± 0.66	−3.39 ± 0.39	0.29 ± 0.67	−0.69 ± 0.43
IRS28 ...	2.00 ± 0.38	−5.29 ± 0.42	2.27 ± 0.35	−5.85 ± 0.31	0.27 ± 0.52	−0.56 ± 0.52
SiO-15 ...	−2.47 ± 0.98	0.77 ± 2.10	−2.18 ± 0.51	−0.56 ± 0.12	0.29 ± 1.10	−1.33 ± 2.10
IRS10EE ...	0.04 ± 0.08	−2.09 ± 0.07	0.73 ± 0.23	−1.92 ± 0.27	0.69 ± 0.24	0.17 ± 0.28
IRS15NE ...	−1.96 ± 0.07	−5.68 ± 0.12	−2.40 ± 0.48	−6.29 ± 0.35	−0.44 ± 0.49	−0.61 ± 0.37
IRS17 ...	−1.61 ± 1.08	−0.75 ± 1.22	0.20 ± 0.59	−1.67 ± 0.62	1.81 ± 1.23	−0.92 ± 1.37
IRS19NW ...	1.19 ± 0.14	−0.43 ± 0.31	−0.60 ± 3.08	−0.54 ± 3.47	−1.79 ± 3.08	−0.11 ± 3.48

Table 4.4: Radio–infrared proper motion alignment. μ_x and μ_y are proper motions relative to SgrA* toward the east and north, respectively. Differenced motions (infrared minus radio) are indicated with the superscript "Dif". Radio motions are in a reference frame tied to SgrA*; infrared motions are relative motions, with an average of ≈ 400 star motions removed.

and, owing to its extreme brightness, the infrared measurements are compromised by detector saturation. Additionally, the radio measurements are subject to significant uncertainty from the large SiO maser shell size. After removing IRS 7, the weighted mean differences between the radio and IR motions change only slightly and become +0.63 ± 0.21 mas/yr toward the east, and −0.32 ± 0.18 mas/yr toward the north.

When comparing how well the IR frame matches the radio frame, we need to consider the statistical uncertainty of the average IR motion, which has been removed. For most epochs, the average motion is based on ≈ 400 stars, each of which has a typical motion of ≈ 100 km/s. Thus, the mean IR motion should have an uncertainty of roughly 100 km/s / $\sqrt{400}$ ≈ 5 km/s. Adopting the result with IRS 7 removed, converting to linear speeds for a distance of 8.0 kpc to the Galactic Center (Reid 1993), and adding in quadrature a ≈ 5 km/s uncertainty for the mean IR motion removed from each coordinate, implies that the infrared stellar cluster moves +24 ± 9 km/s toward the east, and −12 ± 9 km/s toward the north, with respect to SgrA*. The northward component motion does not deviate significantly from zero; the eastward component formally presents a 2.7σ significance. Combining these components in quadrature formally yields a speed difference of 27 ± 9 km/s. However, at this time, we do not consider that we have firmly detected motion of the stellar cluster, and we adopt a 2σ upper limit of 45 km/s for the proper motion of the stellar cusp with respect to SgrA*.

4.5 Enclosed Mass versus Radius from SgrA*

Estimates of the enclosed mass versus projected radius from SgrA*, based on infrared stellar motions, rely on *relative* motions not tied directly to SgrA*. Since, the 3-dimensional motions of the SiO masers in this paper are both very accurate and directly tied to SgrA*, they provide valuable information on the enclosed mass within projected radii of 0.2 to 2 pc of SgrA*.

In Paper II, we derived a lower limit to the enclosed mass at the projected radius of each star, assuming the stellar motions reflect gravitational orbits dominated by a central point mass. Given the 3-dimensional speed, V_{total}, and projected distance from SgrA*, r_{proj}, for each star, we obtained a strict lower limit to the mass enclosed, M_{encl}, within the true radius, r, of that star from SgrA*. For a given enclosed mass, semi-major axis and eccentricity (e), the greatest orbital speed occurs at pericenter for $e \approx 1$. Since the *projected* pericenter distance cannot exceed the true distance, we obtained

$$M_{encl} \geq \frac{V_{total}^2 r_{proj}}{2G} \ . \tag{4.1}$$

Note that this enclosed mass limit is a factor of two lower than would be obtained for a circular orbit. This lower limit approaches an equality only when three criteria are met: 1) $r_{proj} \approx r$, 2) the star is near pericenter, and 3) it has an eccentricity near unity. The *a priori* chance of any star satisfying all three of these criteria is small, especially since a star in a highly eccentric orbit spends very little time near pericenter. Thus, Eq. 4.1 provides a very conservative limit on enclosed mass.

We evaluate the lower limit to M_{encl} using Eq. 1 by adopting conservatively the smallest total velocity allowed by measurement uncertainties, i.e. by subtracting 2σ from V_{total} in Table 4.2.2 before calculating a mass limit. The mass limits, given in Table 4.2.2, are mostly consistent with the enclosed mass versus projected distance from SgrA* given by Genzel et al. (1997) and Ghez et al. (1998). For many of the stars, the lower mass limits are well below the estimated enclosed mass curves, as expected given the very conservative nature of the calculated limits.

Our most significant lower mass limit is from IRS 9. In Paper II, we arrived at a limit $> 4.5 \times 10^6 \, M_\odot$, which exceeded the then favored model of a $2.6 \times 10^6 \, M_\odot$ black hole (Genzel et al. 1997; Ghez et al. 1998), plus a $0.4 \times 10^6 \, M_\odot$ contribution from the central stellar cluster (Genzel et al. 2003a), by about 50%. With our improved proper motions, we now find a more stringent limit of $> 5.1 \times 10^6 \, M_\odot$ at a projected radius of 0.33 pc from SgrA*.

Fig. 4.5 displays our enclosed mass versus radius constraint based on the 3-dimension motion of IRS 9, along with other constraints in the recent literature. The current best estimate for the mass of the SMBH (SgrA*) is $(3.9 \pm 0.2) \times 10^6 \, M_\odot$, for the distance to the Galactic Center, R_0, of 8.0 kpc. This mass estimate comes from an unweighted average of the results of Eisenhauer et al. (2005a) and Ghez et al. (2005a), based on stellar orbit determinations. Adding in a $0.4 \times 10^6 \, M_\odot$ contribution from the central stellar cluster, based on a density profile of $1.2 \times 10^6 \, (r/0.39 \, \text{pc})^{-1.4} \, M_\odot \, \text{pc}^{-3}$ by Genzel et al. (2003a), yields $4.3 \times 10^6 \, M_\odot$, still leaving a discrepancy of $0.8 \times 10^6 \, M_\odot$,

4.5. Enclosed Mass versus Radius from SgrA*

for $R_0 = 8$ kpc. Formally, this is about a 3σ discrepancy, assuming an uncertainty of $\pm 0.2 \times 10^6\ M_\odot$ in the mass estimate of SgrA* and an estimated $\pm 30\%$ uncertainty in the mass of the stellar cusp.

Since we do not know the line-of-sight distance of IRS 9 from SgrA*, one might be tempted to argue that $r \approx 2 r_{proj}$ and the star is simply sensing an enclosed mass of $1.0 \times 10^6\ M_\odot$ from the central stellar cluster at that radius. However, the mass limit derived from Eq. 4.1 scales as r and would be approximately $10^7\ M_\odot$ for $r = 2 r_{proj} = 0.66$ pc. Thus, the mass discrepancy only *increases* for $r > r_{proj}$, as shown by the slanted line in Fig. 4.5 (but see sect. 5.2).

How can the lower limit to the enclosed mass provided by IRS 9 be explained? We now discuss some possibilities.

4.5.1 Dark Matter in the Central Stellar Cluster

One could explain the motion of IRS 9, were the central stellar cluster to contain dark matter (in addition to SgrA*) whose mass exceeds $0.8 \times 10^6\ M_\odot$ within $r = 0.33$ pc. Morris (1993) estimates that $\sim 10^6\ M_\odot$ of "dark" stellar remnants (eg, white dwarfs, neutron stars, black holes) could have accumulated in the inner few tenths of a parsec of the Galaxy. Mouawad et al. (2005) show that, with data available at the time, the orbital fit of star S2 allows for (but does not require) $0.2 \times 10^6\ M_\odot$ of dark matter distributed within 0.001 pc of SgrA*. Should such a dark component exist and extend to greater radii, it might explain some of the IRS 9 mass discrepancy. However, other estimates of the total mass in black holes in the central few tenths of a pc do not exceed $\sim 0.2 \times 10^6\ M_\odot$ (Miralda-Escudé & Gould 2000; Freitag, Amaro-Seoane & Kalogera 2006). Given these estimates and the evidence from other enclosed mass indicators that do not support $\sim 10^6\ M_\odot$ of dark matter within ≈ 1 pc of SgrA* Genzel et al. (2000), it seems unlikely that a dark component could explain more than a modest fraction of the IRS 9 mass discrepancy.

4.5.2 IRS 9 not bound to the central parsec

A critical assumption for calculating the minimum enclosed mass using IRS 9's space velocity (Eq. 1) is that it is in a bound orbit dominated by a central point mass. If IRS 9 is in a highly eccentric orbit with a semimajor axis greater than a few parsecs, this assumption can break down. In such a case, the star's space velocity could exceed the "local" escape velocity, based on the mass enclosed at its current radius, but still be bound at a larger radius. For example, a star could be bound by mass within ≈ 10 pc of SgrA*, but observed plunging into the inner few tenths of a parsec at a speed that

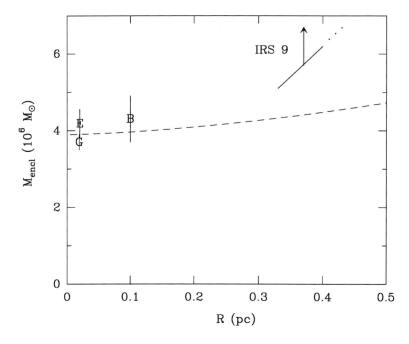

Figure 4.5: Constraints on the enclosed mass as a function of radius (R) from SgrA*. The estimates labeled "E" and "G" are from fitting stellar orbits by Eisenhauer et al. (2005a) and Ghez et al. (2005a), respectively, and the estimate labeled "B" is from a statistical analysis of the "clockwise stellar disk" by Beloborodov et al. (2006). Our lower limit on enclosed mass from the 3-dimensional motion of IRS 9, assuming that it is bound in the region dominated gravitationally by Sgr A* (see sect. 5.2), is indicated by a sloping line and arrow. The minimum distance of IRS 9 from SgrA* is its projected distance of 0.33 pc. The sloping line corresponds to the mass limit for reasonable values of the unknown line-of-sight distance of IRS 9 from SgrA*; this increases both the radius and the mass limit linearly. The dashed line indicates the combined contribution of a point mass of $3.9 \times 10^6 \, M_\odot$ and a central stellar cusp (Genzel et al. 2003a). The uncertainty in the dashed line is dominated by the uncertainty of $\pm 0.2 \times 10^6 \, M_\odot$ in the point-mass, with a smaller contribution of perhaps $\pm 0.1 \times 10^6 \, M_\odot$ from the central stellar cusp at $R = 0.33$ pc. All values assume $R_0 = 8.0$ kpc; the discrepancy between the dashed line and the IRS 9 limit cannot be removed for values of $R_0 < 9$ kpc.

4.5. Enclosed Mass versus Radius from SgrA*

makes it appear unbound. This could happen via gravitational scattering starting either at small radii and increasing orbital energy or at large radii and removing angular momentum. Van Langevelde et al. (1992) suggest a similar explanation for three OH/IR stars with large radial velocities; these stars are seen projected tens of parsecs from SgrA* and would require semimajor axes of a few kpc.

Consider a Galactic Center star with a large semimajor axis and little angular momentum ($e \approx 1$), so that it essentially "falls" toward the Center. We derive the infall velocity for a star that starts "falling" at a radius, r_{max} and reaches a radius of r_0. Assume a central point mass, M_{BH}, plus an extended component with density, $\rho(r)$. The kinetic energy gained by a star falling from r to r_0 is equal to the difference in gravitational potential energy at those radii. Gravitational potential energy per unit mass, U_m, for a spherically symmetric mass distribution has the properties that for mass interior to r

$$U_m = -\frac{G}{r} \int_0^r \rho(r) \, 4\pi r^2 \, dr \, , \quad (4.2)$$

and for mass exterior to r

$$U_m = -G \int_r^{r_{max}} \frac{\rho(r)}{r} \, 4\pi r^2 \, dr \, . \quad (4.3)$$

(For spherically symmetric systems, interior mass acts identically as a point mass equal to the enclosed mass located at the center of the distribution, while exterior mass results in zero gravitational force and a *constant* gravitational potential dependent on its radial position, but independent of the position of a "test" mass.)

Following Genzel et al. (2003a), $\rho(r) = \rho_0 (r/r_0)^\alpha$, where $\rho_0 = 1.2 \times 10^6 \, M_\odot \, pc^{-3}$, $r_0 = 0.39$ pc, and $\alpha = -2.0$ for $r \geq r_0$ pc and $\alpha \approx -1.4$ for $r < r_0$ pc. This leads to an enclosed stellar mass within r_0 of $M_0/1.6$, where $M_0 = 4\pi \rho_0 r_0^3 = 0.9 \times 10^6 \, M_\odot$, plus a contribution of M_{BH} from SgrA*. Adding the contributions to the potential from different mass components for a star at radius r (for $r \geq r_0$) gives

$$U_m = -\frac{G}{r} M_{BH} - \frac{G}{r} \int_0^{r_0} \rho(r) \, 4\pi r^2 \, dr - \frac{G}{r} \int_{r_0}^r \rho(r) \, 4\pi r^2 \, dr - G \int_r^{r_{max}} \frac{\rho(r)}{r} \, 4\pi r^2 \, dr \quad (4.4)$$

The first three terms on the right hand side of Eq. 4.4 sum the effects of the mass components interior to r, and the fourth term is the contribution from the mass exterior to r. Evaluating Eq. 4.4 we find

$$U_m = -\frac{G}{r} M_{BH} - \frac{G M_0}{r \, 1.6} - \frac{G}{r} M_0 \left(\frac{r}{r_0} - 1 \right) - \frac{G}{r_0} M_0 \left(\ln(r_{max}/r_0) - \ln(r/r_0) \right) \quad (4.5)$$

For a star "falling" from r_{max} to r_0, the kinetic energy per unit mass gained is equal to the difference in potential energy per unit mass. From Eq. 4.5, we find

$$\frac{1}{2}v^2 = U_m(r_{max}) - U_m(r_0) \ . \tag{4.6}$$

where

$$U_m(r_{max}) = -\frac{G}{r_{max}}\left(M_{BH} + M_0(\frac{r_{max}}{r_0} - 0.38)\right) \ , \tag{4.7}$$

and

$$U_m(r_0) = -\frac{G}{r_0}\left(M_{BH} + M_0(0.62 + \ln(r_{max}/r_0))\right) \ . \tag{4.8}$$

Evaluating Eq. 4.6 for $r_0 = 0.39$ pc, which is a reasonable value for the 3-dimensional radius of IRS 9, gives $v > 370$ km/s for an initial radius $r_{max} > 9$ pc. Thus, if IRS 9 is in a highly eccentric orbit that takes it out to a radius of > 9 pc, it could achieve its very high observed 3-D velocity without violating the enclosed masses estimated by other methods.

A priori it might seem very unlikely that even one of 15 stars with detectable SiO masers would have such an orbit and be observed near closest approach to SgrA* (where it spends little time). However, it is beyond the scope of this paper to evaluate the likelihood, especially with the limited statistics available at this time.

4.5.3 R_0 exceeds 9 kpc

Were the mass of SgrA* $> 4.7 \times 10^6 \ M_\odot$, no mass discrepancy would exist. The best current mass estimates are based on fitting orbits for many stars and should be robust. However, the greatest uncertainty in the mass of SgrA* comes its strong dependence on the adopted value of $R_0 = 8.0$ kpc for the distance to the Galactic Center. Eisenhauer et al. (2005a) derive central masses from orbit fitting of $4.06 \times 10^6 \ M_\odot$ when adopting $R_0 = 8.0$ kpc and $3.61 \times 10^6 \ M_\odot$ for a best fit $R_0 = 7.62$ kpc. These values suggest an enclosed mass $M_{encl} \propto R_0^{2.4}$. Our mass limit based on IRS 9's 3-D motion would also increase with R_0, but more weakly. Since the LSR velocity is the dominant component in the 3-dimensional motion for IRS 9 (and is not dependent on R_0), our minimum mass estimate (Eq. 1) scales approximately as $M_{encl} \propto R_0^{1.3}$, mostly through r_{proj}. Allowing R_0 to increase to about 9 kpc removes the mass discrepancy. However, such a large value for R_0 seems very unlikely (Reid 1993; Eisenhauer et al. 2005a).

4.5. Enclosed Mass versus Radius from SgrA*

4.5.4 Non-zero V_{LSR} for SgrA*

Were SgrA* moving toward the Sun along the line-of-sight with a speed > 30 km/s, this would lower V_{total} and, hence, the M_{encl} limit to $> 4.4 \times 10^6 \, M_\odot$. While it seems very unlikely that a super-massive object would have such a large motion (Reid et al. 2003), we now consider this possibility. One method to approach this problem is to average the velocities of large samples of stars very close to SgrA*, assuming that this average would apply to SgrA*.

Our sample of SiO maser stars, which should be nearly complete in the LSR velocity range -131 to $+95$ km/s, does not show any obvious bias. Infrared observations of CO band-head velocities from late-type stars in the central parsecs yield average LSR velocities that are not statistically different from zero. For example, the integrated CO-band head velocities (within a 20" diameter aperture) of McGinn et al. (1989) indicate positive (negative) velocities at positive (negative) Galactic longitude, consistent with the direction of Galactic rotation, and a value of -10 ± 25 km/s at the position of SgrA*. (However, these authors find possibly significant stellar velocities of -47 ± 8 km/s for four pointing offsets *perpendicular* to the Galactic plane.) Individual stellar velocities compiled by Rieke & Rieke (1988) of 54 stars projected within ≈ 6 pc of SgrA* have a mean velocity of -20 ± 11 km/s. Alternatively, Winnberg et al. (1985) and Sjouwerman et al. (1998) measured velocities of OH masers for 33 and 229 OH/IR stars, respectively, within about ≈ 40 pc of SgrA*, which yield average velocities of $+7 \pm 11$ and $+4 \pm 5$ km/s. Overall, it appears that radial velocities of stars near SgrA* suggest an average LSR velocity near zero, within ≈ 20 km/s.

4.5.5 IRS 9 is (or was) in a binary

Were IRS 9 in a tight, massive binary, perhaps a significant portion of its space velocity might be contributed by internal orbital motion, possibly reducing its speed with respect to SgrA*. However, we have observed IRS 9 for about 8 years and see no changes in its radial or proper motion velocity components. The spectra of IRS 9 shown in Fig. 4.2 are characteristic of Mira variables, which show variable emission over a range of 5 to 10 km/s about the stellar velocity. We estimate that the stellar radial velocity of IRS 9 has changed by less than 2 km/s over 8 years. Also, the proper motion components are well-fit by constant velocities, with 2σ upper limits to accelerations of 0.4 and 0.6 mas y^{-2} (15 and 23 km/s y^{-1} at 8.0 kpc) for the eastward and northward components, respectively.

The observed changes in radial velocity (or proper motion components) would have different characteristics depending on the relative values of the time span of the observations, Δt, and the orbital period, P. For $\Delta t \geq P/4$, we would have sampled large changes in the orbital mean anomaly and hence would have seen quasi-random changes of magnitude equal to the orbital speed. Our limit of < 2 km/s change in the radial velocity over 8 y, would place a limit of ≈ 2 km/s for the radial component of any orbital velocity. Alternatively, if $\Delta t < P/4$, then we could be sampling only a small portion of an orbit and detecting a velocity change might be difficult. However, setting $\Delta t = 8$ y requires $P > 32$ y. Since any bright companion for IRS 9 would have been observed, we adopt companion mass m of $\leq 10\ M_\odot$. This would even allow for most black hole companions. For $m < 10\ M_\odot$ and $P > 32$ y, we find an upper limit for an orbital speed of < 20 km/s. Thus, a binary orbital contribution to the observed space velocity of IRS 9 could not exceed ≈ 20 km/s and likely would be considerably less. Thus, it is highly unlikely that the extreme velocity of IRS 9 could be explained as owing to a binary orbit.

Could IRS 9 have been in a binary system and come unbound (or bound at a much larger radius as discussed in sect. 5.2) after a close encounter with SgrA*? A small number of "hyper-velocity" stars are thought to have been ejected from the Galactic Center in this manner (Hills 1988; Yu & Tremaine 2003; Brown et al. 2005). However, these are estimated to be very rare events (< 1 in 10^5 y) and we are statistically unlikely to be witnessing a newly created hyper-velocity star so close to SgrA*. All hyper-velocity stars discovered to date are early-type main-sequence stars; they are found in the outer Galaxy and are moving at speeds of ~ 1000 km/s, even after climbing out of the gravitational potential of the inner Galaxy. Main sequence stars can survive the strong tidal forces experienced during close encounters with SgrA*. However, IRS 9 is an AGB star and, thus, is a very extended (≈ 300 solar radii) and low surface-gravity object. It is unclear if such a star could survive the ejection event, without losing its extended atmosphere.

4.6 Conclusions

We have now measured the radio positions and proper motions of 15 late type stars with SiO maser emission in the Galactic Center stellar cluster. All but two of these stars have been detected at three or more epochs and have measurement accuracies of ≈ 1 mas in position and ≈ 0.3 mas/yr in proper motion. Nine of these stars have multi-epoch measurements of proper motions at infrared wavelengths. A comparison of the radio motions, which are relative to SgrA*, with the infrared motions indicates that the stellar cusp moves with SgrA* to within 45 km/s.

The three-dimensional speeds and projected distances of individual stars from SgrA*

4.6. Conclusions

yield lower limits to the enclosed mass. The enclosed mass limit for one star, IRS 9, exceeds current estimates of the combined mass of SgrA* and the luminous stars in the cusp within the central parsec. This result is puzzling, but might be explained, for example, by a combination of (i) a population of dark stellar remnants in the central parsec, (ii) IRS 9 being on a plunging "orbit" with a semimajor axis $\gg 1$ pc, and/or (iii) a value of $R_0 > 8$ kpc.

Acknowledgements: We thank S. Gillessen for comments on the paper, and A. Loeb and R. O'Leary for discussions on the extreme motion of IRS 9.

Chapter 5

Kinematics of the CO star cluster

Original publication: S. Trippe, S. Gillessen, H.L. Maness, F. Martins, O.E. Gerhard, T. Ott, F. Eisenhauer, S. Rank, K. Dodds-Eden and R. Genzel 2008, *The kinematics of the Galactic Center CO star cluster*, in preparation

Abstract: We aim at a detailed description of the kinematic properties of the late-type CO absorption star population among the Galactic Center (GC) cluster stars. This cluster is composed of a central supermassive black hole (SgrA*) and a self-gravitating system of stars. Understanding its kinematics thus offers the opportunity to understand the dynamical interaction between a central point mass and the surrounding stars in general, especially in view of other galactic nuclei.

We use AO assisted near-infrared imaging and integral field spectroscopy using the instruments NAOS/CONICA and SINFONI at the VLT. We obtain proper motions for about 5500 stars, 3D velocities for about 660 stars, and acceleration limits (in the sky plane) for about 750 stars. Global kinematic properties are analyzed using velocity and velocity dispersion distributions, phase-space maps, and two-point correlation functions.

We detect for the first time significant cluster rotation in the sense of the general Galactic rotation in proper motions. Out of the 3D velocity dispersion, we derive an improved statistical parallax for the GC of $R_0 = 8.37 \pm 0.29$ kpc. The stellar 3D speeds follow a Maxwellian distribution. We do not find stars whose kinematics would require non-Gaussian processes (like star ejection by three-body interactions). We find an upper limit of 2% for the amplitude of fluctuations in the phase-space distribution of the cluster stars compared to a uniform, spherical model cluster. Using upper limits on accelerations, we constrain the minimum line-of-sight extensions of observed star ensembles within the innermost few (projected) arcsec. The stars within 0.7" radius from the star group IRS13E do not co-move with this group, making it unlikely that

IRS13E forms a substantial cluster. In total, the GC late-type cluster is well described as a uniform, dynamically relaxed, phase-mixed system.

5.1 Introduction

The dynamical properties of the Galactic Center (GC) star cluster, which hosts the radio source and supermassive black hole (SMBH) SgrA*, have been subject to intensive research for more than a decade. Since due to strong interstellar extinction ($A_V \sim 30$) the GC stars can be observed only in the infrared (IR), most of the work in this field is based on near-infrared (NIR) data ranging from H to L bands (1.5 – 4 µm; see Fig. 5.1 for an example).

Initially, the central question of this research was whether the GC indeed hosts a central SMBH as had been expected by Lynden-Bell & Rees (1971) even before the discovery of the radio point source SgrA* (Balick & Brown 1974). Based on statistical arguments using the observed velocity dispersions, it was possible to show in the late 1980s and 1990s that a central pointlike mass of a few million solar masses was present. Additionally, increasingly better estimates of the distance to the GC became possible (McGinn et al. 1989 ; Krabbe et al. 1995; Eckart & Genzel 1997; Ghez et al. 1998; Genzel et al. 1996, 1997, 2000).

With improved data quality, due especially to the establishment of speckle imaging and adapive optics (AO) assisted imaging and spectroscopy, as well as longer observation time lines, more direct tests of the central mass became possible. These efforts culminated in the observation of Keplerian star orbits in the immediate vicinity ($\sim$0.5" or $\sim$4000 AU) of SgrA* which allowed a direct geometric determination of the mass $M_\bullet$ of and the distance R_0 to the central SMBH (Schödel et al. 2002, 2003; Ghez et al. 2003, 2005a; Eisenhauer et al. 2003a, 2005a). Following Eisenhauer et al. (2005a), we adopt throughout this paper a canonical distance $R_0 = 8$ kpc and a distance-scaled mass $M_\bullet = (4.1 \pm 0.4) \cdot 10^6 \cdot (R_0/8\text{kpc})^{2.3}$ M$_\odot$. For the later discussion one should keep in mind that for this GC distance the image scale is 1 arcsec $\simeq$ 39 mpc $\simeq$ 8000 AU in position and 1 mas/yr $\simeq$ 38 km/s in velocity.

In addition to these advances, a dynamically complex structure of the central cluster on scales of $\sim$1–10" became visible. It was possible to show that the cluster is composed of two main populations: a population of dynamically relaxed, evolved, old (several Gyr) isotropically distributed late-type CO absorption line stars, and a relatively small population of young ($\sim$6 Myr) OB- and Wolf-Rayet-stars, located in the central arcsecond and in two disks centered on SgrA* (Genzel et al. 2003a; Paumard et al. 2006; Maness et al. 2007).

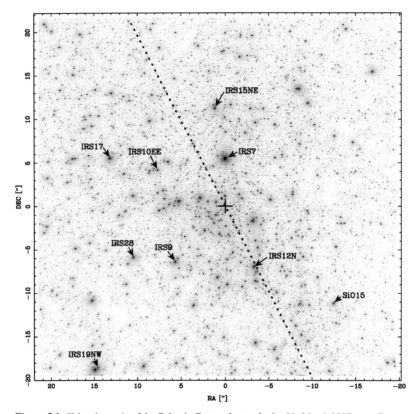

Figure 5.1: K-band mosaic of the Galactic Center cluster obtained in March 2007; coordinates are RA and DEC in arcsec relative to SgrA*. This image covers a FOV of $\sim 1.5 \times 1.5$ pc. The contrast was enhanced by applying unsharp masking. The position of SgrA* is marked with a cross in the image center. The dotted line crossing the image indicates the Galactic plane. Stars with names are SiO maser stars used for defining an astrometric reference frame.

As most, if not all, nearby galactic systems contain central SMBHs (e.g. Ferrarese & Ford 2005, and references therein), analyzing the Galactic Center system enhances the understanding of galaxy cores in general. A nice example is the case of M31 and its triple nucleus, where a $\sim 10^8$ $M_\odot$ black hole drives the complex dynamics of the system (Kormendy & Bender 1999; Bender et al. 2005).

In this article we focus on the properties of the population of evolved late type, CO absorption stars. We present the as yet most precise kinematical analysis of the central star cluster. This work is based on proper motions and radial velocities extracted from diffraction limited imaging and spectroscopy data obtained from 2002 to 2007.

This paper is organized as follows. In section 2, we summarize the data acquisition and reduction. Section 3 describes the extraction of stellar positions and proper motions from imaging data. Section 4 gives an overview on the collection of line-of-sight velocities from integral-field spectroscopy data. In section 5, we present our findings and discuss them. Section 6 summarizes our results and conclusions.

5.2 Observations and data reduction

This work is based on observations with the 8.2-m-UT4 (Yepun) of the ESO-VLT on Cerro Paranal, Chile. For obtaining *imaging* data we used the detector system NAOS/CONICA (NACO for short) consisting of the AO system NAOS (Rousset et al. 2003) and the 1024×1024-pixel NIR camera CONICA (Hartung et al. 2003b).

We obtained 10 data sets in H and K bands with a pixel scale of 27 mas/pixel (large scale) covering 6 epochs (May 2002, May 2003, June 2004, May 2005, April 2006, March 2007). In this mode each image covers a field of view (FOV) of 28×28". During each observation the camera pointing was shifted so that the typical FOV of an entire data set is around 40×40", centered on SgrA*.

We executed a much larger number of observations using a smaller pixel scale of 13 mas/pixel (small scale), thus resulting in an image FOV of 14×14" and, again using shifted pointings, typical observation FOVs of 20×20". In total we obtained 42 H and K band image sets, 5 to 10 per year in roughly monthly sampling.

To all images we applied sky-subtraction, bad-pixel and flat-field correction. In order to obtain the best possible signal-to-noise ratios and maximum FOV coverages in single maps, we combined all good-quality images obtained in the same night into mosaics.

For obtaining *spectroscopic* data we used SINFONI, a combination of the integral field spectrometer SPIFFI (Eisenhauer et al. 2003b,c) and the adaptive optics system MACAO (Bonnet et al. 2003, 2004). The data output is structured as data cubes with two spatial axes with dimensions of 64 and 32 pixels respectively, and one spectral axis of 2048 pixels length. Thus SINFONI provides diffraction-limited 64×32 pixel images with a spectrum for each image pixel.

Depending on the plate scale, individual cubes covered regions of 0.8 × 0.8", 3.2 × 3.2", or 8 × 8"; the last scale was used in seeing-limited mode only. The spectra covered either the K band (with a spectral resolution of $R = 4500$) or the band range H+K ($R = 2500$).

After sky subtraction, bad-pixel-, and flat-field-correction, each spectrum was calibrated spectrally with an emission line gas lamp. Atmospheric absorption features were removed by dividing by the spectrum of a calibration star.

5.3 Astrometry

5.3.1 The procedure

From our imaging data we extracted time-resolved stellar positions. The way from individual images to astrometric positions is outlined in this section.

In part one of the data processing we construct one mosaic out of the ∼20...100 individual frames obtained in a given observation night. This requires the following steps:

1. Computing the parameters of instrumental geometric distortion for the given data set (distortion correction).

2. Obtaining the offsets and rotation angles between the individual exposures (image registration).

3. Mapping all individual exposures into one image grid (mosaicking).

Step 1 we outline in subsection 3.2; steps 2, 3 we describe in subsection 3.3.

Part two of the data processing covers the way from star positions in a mosaic to absolute astrometric coordinates:

1. In each mosaic, extract star positions in image coordinates (i.e. in units of pixels).

2. Define one mosaic as reference image (or zero-image).

3. For the reference image, extract image positions of nine SiO masers stars with well-known absolute astrometric coordinates measured in radio. Compute the parameters of the transformation reference image → absolute coordinates.

5.3. Astrometry

4. Apply the coordinate transformation to all star positions extracted from the reference image. Pick a set of ∼560 stars serving as reference ensemble ("cluster frame").

5. For each mosaic, transform image coordinates into absolute coordinates using the ∼560 reference stars. Image coordinates of the refence stars are extracted from the respective mosaic. Absolute coordinates are the radio coordinates obtained from the reference image. This procedure assumes that the reference ensemble is stationary in average.

6. Extract proper motions by fitting star coordinates vs. time with linear functions.

Steps 1–5 we address in subsection 3.4, step 6 we describe in subsection 3.5.

5.3.2 Geometric distortion

In order to avoid systematic alignment errors when mosaicking single images, we first corrected the individual frames for the geometric distortion of the CONICA imager. As there is no publicly available description of the instrumental distortion properties of NACO, we extracted the necessary parameters from our data by executing the following steps:

1. Combination of all individual frames to be mosaicked via simple shift-and-add (SSA) with integer-pixel accuracy.

2. Constructing a list of many (∼200) good (meaning bright, but unsaturated stars well separated from neighbouring sources) reference stars distributed over the entire FOV. For source selection, the SSA image is used.

3. Re-identification of all reference stars located in the FOV of each individual image, followed by determining their positions.

4. Computation of star separations for all stars in each image. This results in a net of baselines for each image. Baselines present in more than one image are subject to inter-image comparison.

5. Modelling the distortion correction using the radially symmetric standard ansatz

$$\mathbf{r} = \mathbf{r}'(1 - \beta \mathbf{r}'^2)$$

with

$$\mathbf{r} = \mathbf{x} - \mathbf{x}_C \quad \text{and} \quad \mathbf{r}' = \mathbf{x}' - \mathbf{x}_C$$

(e.g. Jähne 2005)[1]. Here $\mathbf{x}$ and $\mathbf{x}'$ are the true and distorted image coordinates respectively, β is a parameter describing the strength of the grid curvature, and $\mathbf{x}_C$ is the zero point of the distortion on the detector.

The three model parameters (β, x_C, y_C) were fit by a χ^2 minimization. Geometric distortion implies that the detector plate scale is a function of the detector position; thus the length of a given baseline (in units of pixels) depends on its location on the detector. The optimum parameter set is found by iteratively comparing all baselines in all images shifted on sky, applying the temporary parameters to the reference star detector coordinates, and checking for the improvement.

In case of the *large scale* (27 mas/pixel) images the parameter fits could be executed in a straight forward manner. We made use of the analytic fit engine *FindMinimum* implemented in the software package *Mathematica*[2].

The results found were located in the ranges

$x_C \simeq 577 \ldots 629$ pixels
$y_C \simeq 775 \ldots 823$ pixels
$\beta \simeq 2.97 \ldots 3.40 \cdot 10^{-9}$ pixels^{-2}

For the *small scale* (13 mas/pixel) data sets this procedure was not applicable. Due to a less significant distortion and the smaller number of reference stars available ($\sim$100), the analytic fit algorithm usually did not converge towards a reliable result. Thus we constructed a stochastic minimization algorithm which searches the parameter space iteratively using the following scheme:

1. Compute the value of the cost function (i.e. the function to be minimized) at the actual position in parameter space.

2. Select a second position in parameter space and compute the value of the cost function at that position.

3. If the value of the cost function at the new position is smaller than the actual one: move there. Otherwise: stay at present position.

---

[1] See also the electronic manual of the public Gemini North Galactic Center Demonstration Science Data Set for an application on GC imaging data.
[2] Wolfram Research, Inc., Champaign, IL, USA

5.3. Astrometry

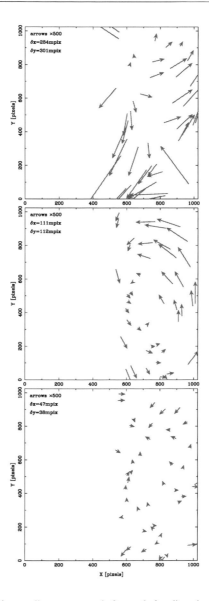

Figure 5.2: Residual image alignment errors before and after distortion correction and image registration, comparing the overlap area (right hand half of image no. 1) of two 27 mas/pixel scale K band images obtained in March 2007. Arrows mark absolute values (1 unit = 2 millipixels) and directions of residuals, δx (δy) are the rms of residuals in x (y). *Top panel:* Subpixel accurate shift-and-add only. *Central panel:* After correcting for geometric distortion, before registration. *Bottom panel:* After distortion correction and registration.

4. Repeat steps 1–3 until a fixed number of iterations is completed.

Starting from a given initial position in parameter space, in iteration n for each parameter p a new value (step 2) is computed as

$$p_{n+1} = p_n + s(n) \cdot (1/z_{[0,1]} - 1) \cdot \varepsilon(0.5 - z_{[0,1]})$$

with

$$s(n) = s_0 \cdot 10^{-n/N}$$

Here N is the maximum number of iterations, $z_{[0,1]}$ a random number in the range $[0,1]$, $\varepsilon(x)$ the sign function returning -1 or $+1$ depending on the sign of x, and s_0 the initial step size.

This definition assures that (1) the algorithm cannot be easily trapped in a local minimum, as even extreme search radii are occasionally tested, and (2) the vicinity of the best-so-far-solution found at the end of the search time is explored with reasonable accuracy, as the average search radius decreases exponentially with time (thus increasing the "selection pressure" on the algorithm). The idea for this definition was taken from the concept of Simulated Annealing introduced by Kirkpatrick et al. (1983). The distortion parameters were in the ranges

$$x_C \simeq 573 \ldots 839 \text{ pixels}$$
$$y_C \simeq 629 \ldots 948 \text{ pixels}$$
$$\beta \simeq 2.06 \ldots 13.27 \cdot 10^{-10} \text{ pixels}^{-2}$$

One has to note, that stochastic minimization algorithms like the one described above in general find a *good* solution for a given problem, not the *optimum* solution. Thus one can find a wide range of solutions even if there is no evolution of the tested parameters, as it appears to be the case here (especially for β). In spite of this limitation, it is possible to state that for the large scale images ($\beta \sim 3 \cdot 10^{-9}$ pixels^{-2}) the distortion is clearly stronger than for the small scale images ($\beta \sim 5 \cdot 10^{-10}$ pixels^{-2}).

5.3.3 Image registration and mosaicking

After extracting the distortion parameters, all single frames were registered with respect to a common coordinate grid to ensure a sub-pixel accurate alignment. To compute the necessary transformations between the single images and the mosaic grid, we used the reference star coordinates and corrected them for distortion. Assuming that

5.3. Astrometry

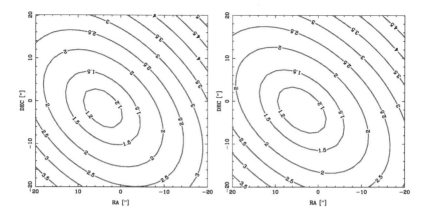

Figure 5.3: Statistical uncertainty of absolute radio reference frame coordinates as a function of position due to transformation errors, separately for RA (*left*) and DEC (*right*). The contour values are given in units of milli-arcseconds. Across the field of view the errors vary in the range 1.2 ... 5 mas. The contours mirror the alignment of the reference stars in the sky plane (see Fig. 5.1).

after the correction of the geometric distortion no systematic higher-order effects are left (meaning the image plate scales are not functions of the positions), the image registration can be described by spatial 1st-order transformations

$$x' = a_0 + a_1 \cdot x + a_2 \cdot y$$
$$y' = b_0 + b_1 \cdot x + b_2 \cdot y$$

which cover translations, rotations, scalings, and shears. These transformations were applied to the undistorted reference star coordinates by (a) defining one master image (usually the first image of a set) as zero point, and (b) computing for each image the parameters of the transformation to the master image coordinates.

The final mosaic was constructed by computing for each image pixel its new position in the mosaic grid. Its flux value was then interpolated into the corresponding mosaic pixels.

Comparing the typical residual alignment errors before and after distortion correction and registration (Fig. 5.2) shows the improvement in mosaic quality clearly. While simple shift-and-add leads to inaccuracies as high as some tenths of a pixel, the distortion correction already means a strong improvement. When taking into account the grid curvature, systematic 1st-order effects (shifts and rotations) are left. After registering the images, the typical residual errors (pairwise between image overlap areas) are of the order ~0.05 pixels, corresponding to ~1.4 mas (large scale) and ~0.7 mas (small scale), respectively.

5.3.4 Positions and coordinates

In order to determine positions and proper motions for as many stars as possible, we first constructed a master source list from a high-quality large scale K band mosaic with ~ 40 × 40" FOV obtained in May 2005. In this image we identified and listed all stars above a given significance threshold using the algorithm *FIND* (Stetson 1987). This algorithm searches an image for positive brightness perturbations and identifies them as stars, if their sharpness and roundness parameters are located within given limits.

Out of the list of all detected sources we excluded those overlapping (i.e. separated by less than ~2 FWHMs / ~130 mas) with neighbouring stars and thus unusable for precise astrometry. The list of remaining "good" stars contains ~6000 objects down to magnitudes of K~18. For all sources the diffraction-limited cores were fit as 2-dimensional elliptical Gaussian brightness distributions.

In order to convert the image positions of the master list stars into radio frame coordinates coordinates, we used a set of 9 SiO maser stars located in the FOV as reference. For these stars global positions are known from radio observations (Reid et al. 2007). Radio positions and NIR image positions were tied by a 1st-order transformation (see sect. 3.2 for comparison). Hereafter, absolute (radio reference frame) coordinates are distances from SgrA* in RA and DEC in units of arcseconds.

As the maser stars show finite statistical errors in both radio and NIR image positions, we examined the influence of these uncertainties on the transformation accuracy using a Monte-Carlo test. We executed 10^5 coordinate transformations, each time using sets of positions with random displacements. The displacements followed Gaussian distributions according to the individual statistical errors. By sampling a coordinate grid with the typical FOV size (positions ±20" from SgrA*) we mapped the transformation uncertainty as a function of position. The results are shown in Fig. 5.3. The contours mirror the geometry of the alignment of the reference stars in the plane of the sky (see also Fig. 5.1); the errors vary in the range 1.2 ... 5 mas. Of particular interest is the accuracy of the global position (0,0) which corresponds to the location of SgrA*, the dynamical center of the GC star cluster. Here the errors were

δRA = 1.26 mas
δDEC = 1.20 mas

5.3. Astrometry

In each individual mosaic the master list stars were re-identified and their detector positions were fit with 2-dim elliptical Gaussian profiles. The formal detector position accuracies were typically ∼0.025 pixels (per coordinate) in both plate scales, i.e. ∼0.68 (0.33) mas in the large (small) plate scale.

Unfortunately, in our NIR data no absolute astrometric reference source is available and the SiO maser stars are present only in large scale (27 mas/pix) images. We therefore defined a relative astrometric reference frame tied to an ensemble of about 560 well-behaved (meaning bright and well separated from neighbouring sources) stars. Due to the slightly different FOVs of the individual data sets, the number of stars actually usable for a given mosaic is somewhat smaller; the average value is 433.

Since typically only ∼100 of these stars are present in the small scale images, we first analyzed the large scale images and computed proper motions for all stars. These proper motions were used to compute the expected astrometric positions of the reference stars for each epoch of a small scale image. The (1st-order) transformation NIR image positions → radio coordinates then uses the expected astrometric positions. This procedure ensures that the small scale images are tied to the reference frame of the large scale images.

5.3.5 Proper motions

Stellar proper motions were computed by fitting linear functions to star positions vs. time. In order to determine proper errors for the stellar velocities, we applied outlier rejection and error rescaling to the data. The typical measurement error of a star position is ∼0.9 mas for both image scales. The timeline of observations is five years. The number of epochs is 10 for the large (27 mas/pix) scale and 42 for the small (13 mas/pix) scale imaging data.

In total we were able to extract proper motions for ∼5500 stars located in the large scale fields; out of these, ∼750 sources were additionally covered by the small scale images. Typical statistical proper motion accuracies are ∼0.18 mas/yr per coordinate for the large scale data sets and ∼0.1 mas/yr per coordinate for the small scale fields. This corresponds to ∼6.8 km/s and ∼3.8 km/s, respectively, for a distance to the GC of 8 kpc. The error distributions are presented in Fig. 5.4.

An additional, systematic uncertainty is introduced by the relative astrometric reference frame. This frame is based on stars with proper motions known only a posteriori and with respect to the star cluster. Thus a systematic motion of the reference frame is possible. Using the average number of applicable large scale image reference stars, which is 433, and the rms velocity of the reference stars (3.6 mas/yr or 137 km/s), we

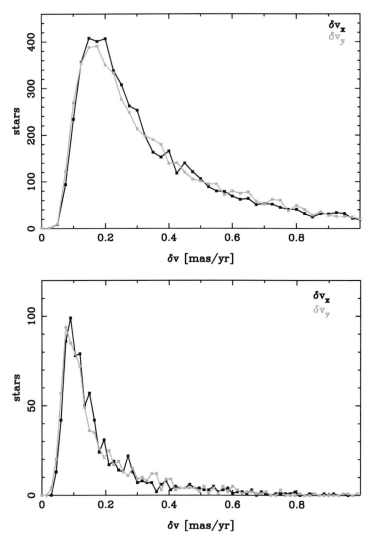

Figure 5.4: Histograms of proper motion errors. *Top panel:* Large scale field. *Bottom panel:* Small scale field. The distributions peak at ∼0.18 mas/yr in case of the large scale and at ∼0.1 mas/yr in case of the small scale, corresponding to 6.8 km/s and 3.8 km/s respectively.

5.4. Spectroscopy

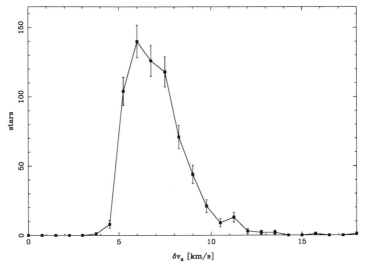

Figure 5.5: Histogram of statistical errors in line-of-sight velocity for about 660 stars. Errors are Poisson $\sqrt{N}$ errors. The distribution peaks at $\sim$7 km/s.

estimate this systematic uncertainty (standard error) to be 0.17 mas/yr or 6.4 km/s.

5.4 Spectroscopy

Stellar spectra were extracted from SINFONI data cubes. For each star the respective source pixels were selected by hand. In a second step surrounding pixels containing the respective backgound spectra were selected. This was done in order to take into account incomplete sky subtraction and flux spillover from neighbouring sources. A corrected star spectrum was then created by subtracting the average of the background pixels from the average of the source pixels.

For further analysis all spectra were normalised to continuum = 1. Radial velocities were extracted by correlating the normalised spectra g_s with a normalised template spectrum g_t. The model spectrum of a CO star obtained from the MARCS stellar model-atmosphere and flux library (Gustafsson et al. 2003) served as template. Main model parameters were temperature $T_{\text{eff}} = 4250$ K, gravitational acceleration $\log g = 0$ [cm/s^2], micro-turbulence velocity $v_{\text{turb}} = 2$ km/s, and solar metallicities. As we were especially interested in the behaviour of late-type stars, we focused our analysis on the CO bandhead lines in the wavelength range $\sim 2.28...2.37$ μm, with the precise cutoff values depending on the individual data quality.

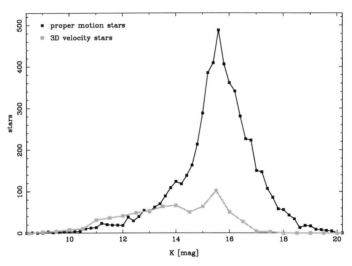

Figure 5.6: K magnitude distributions of all ~5500 proper motion stars (black curve) and the ~660 3D velocity stars (grey curve). Please note the different binsizes. The distributions peak in the range K~15...16; this is the regime of the Red Clump stars. For K>16, the completenesses quickly decrease.

In order to compute a radial velocity for a given star, the template spectrum was interpolated into velocity-shifted (i.e red- or blue-shifted) wavelength frames. For a given velocity bin i with the interpolated template $g_t^{i,j}$ the correlation coefficient

$$c^i = \frac{\sum_j (g_s^j - 1) \cdot (g_t^{i,j} - 1)}{\sqrt{\sum_j (g_s^j - 1)^2 \cdot \sum_j (g_t^{i,j} - 1)^2}} \quad (5.1)$$

was computed; here $j = 1, 2, ..., N$ is the index of the N data points belonging to a spectrum. By successively scanning several thousand bins – typically in ranges -1000 ... +1000 km/s using binsizes of 0.5 or 1 km/s – this algorithm builds up a curve of the correlation as a function of radial velocity. The position of the curve's maximum corresponds to the radial velocity of the examined star. If the maximum correlation was lower than 0.55, the computed velocity was rejected as unreliable. All velocities were corrected for the motion of the local standard of rest.

In total we extracted radial velocities for about 660 late-type stars. Typical (statistical) velocity accuracies are ~7 km/s; their distribution is presented in Fig. 5.5.

5.5. Results and discussion

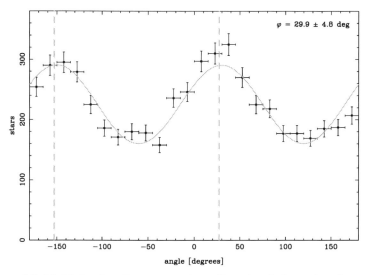

Figure 5.7: Modulations in stellar proper motions. Shown here is the number of stars vs. angle between RA and DEC components of the proper motion vectors. The angle is defined as atan2(v_x, v_y) and counted from north to east. Black points with error bars are the data; horizontal error bars mark the full bin widths, vertical error bars are the Poisson errors. This diagram tests the preferential orientations of proper motion vectors on sky. The vertical dashed grey lines mark the location of the Galactic plane (+27.1° and counter-direction). A cosine fit to the data (grey curve) finds a phase $\phi = +30 \pm 5°$, thus in agreement with the orientation of the Galaxy.

5.5 Results and discussion

5.5.1 Global rotation

Having proper motions at hand for about 5500 stars and 3D-velocity vectors for about 660 stars, we exploited this large data set to extract the dynamical properties of the cluster. Fig. 5.6 shows the K magnitude distribution of the two ensembles. Our data provide information for projected distances from SgrA* up to about 25 arcsec. As a first step we computed velocity dispersions along all coordinate directions.

In order to calculate these parameters we used the method by Hargreaves et al. (1994). This algorithm computes average velocity $\langle v \rangle$ and velocity dispersion σ for a given ensemble of stars using the iterative scheme

$$\langle v \rangle = \frac{\sum_i w_i \cdot v_i}{\sum_i w_i} \qquad (5.2)$$

$$\sigma^2 = \frac{\sum_i [(v_i - \langle v \rangle)^2 - \delta_i^2] \cdot w_i^2}{\sum_i w_i^2} \qquad (5.3)$$

with v_i being the velocity of star i, δ_i being the respective error, and $w_i = 1/(\delta_i^2 + \sigma^2)$ being the star's weight. In general, not more than three iterations are necessary to obtain stable results; we usually used five. We focused on the behaviour of the late type population of stars which is expected to be dynamically relaxed (Genzel et al. 2003a). Therefore we excluded about 100 spectroscopically identified early type stars which are known to mainly move in disks (Paumard et al. 2006). One should however note that probably some more, so far unidentified, early type stars are still included in our proper motion sample.

For the two velocity dispersions in RA (labeled x) and DEC (labeled y) using all ~5500 proper motion stars we found the values

$\sigma_x = 2.668 \pm 0.027$ mas/yr
$\sigma_y = 2.824 \pm 0.028$ mas/yr

implying that the dispersions in RA and DEC are signicantly different (by about 4σ).

To check the amount and geometric structure of a possible anisotropy in the proper motion vectors we tested their preferential orientations on sky. For each star we computed the angle $\psi = \mathrm{atan2}(v_x, v_y)$ which is counted from north to east. The resulting histogram is shown in Fig. 5.7. In case of isotropy the distribution would be flat. The histogram, however, shows a highly significant cosine-like pattern. This pattern is consistent with the signature of a rotating disk seen edge-on, but also with an intrinsic anisotropy in random motions. Fitting this pattern with a cosine profile reveals a phase of $+30 \pm 5°$, which is in agreement with the plane of the Milky Way located at $+27.1°$ (J2000).

In order to translate the distribution shown in Fig. 5.7 into a modulation in proper motions, we used the following ansatz: for a given principal coordinate axis we computed the velocity dispersions parallel ($\sigma_\parallel$) and perpendicular ($\sigma_\perp$) to this axis using all available proper motions. Then we calculated the difference in squares of these two dispersions, $\Delta\sigma^2 = \sigma_\parallel^2 - \sigma_\perp^2$. By rotating the principal axis stepwise on sky, $\Delta\sigma^2$ as a function of the angle is obtained. The resulting curve, here using a step size of $5°$, is shown in the top panel of Fig. 5.8. As always the same set of proper motions is used, the data points are correlated. Using a cosine fit to describe the data, we find an amplitude of $\Delta\sigma_{\mathrm{max}}^2 = 2.00 \pm 0.21$ (mas/yr)^2.

5.5. Results and discussion

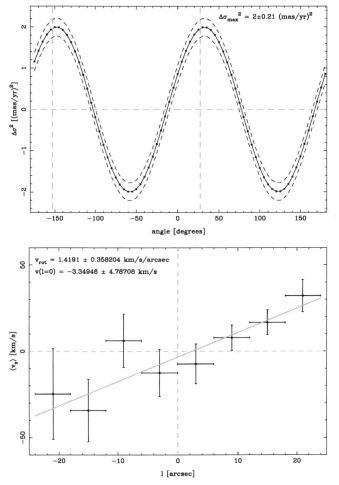

Figure 5.8: Signatures of rotation in proper motions and radial velocities. *Top panel:* Angle on sky (counted from N to E) vs. difference in square dispersions $\Delta\sigma^2 = \sigma_\parallel^2 - \sigma_\perp^2$. $\sigma_\parallel$, $\sigma_\perp$ are the velocity dispersions parallel and perpendicular to a given principal axis which is rotated stepwise. The black dots and the black curve show the observed modulation, dashed black curves mark the 1-σ uncertainty range. Vertical grey dashed lines mark the position of the Galactic plane, the horizontal grey dashed line is the zero level of $\Delta\sigma^2$. The modulation has an amplitude of $\Delta\sigma_{max}^2 = 2.00 \pm 0.21$ (mas/yr)². *Bottom panel:* Average radial velocities $\langle v_z \rangle$ vs. Galactic longitude l. Black points are the data, error bars along the l axis mark the full bin sizes, error bars in velocity direction are 1-σ-errors. Grey dashed lines mark the zero levels of l and $\langle v_z \rangle$. A linear fit to the data (continuous grey line) obtains a rotation velocity of 1.4±0.4 km/s/arcsec.

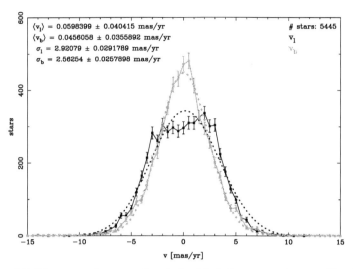

Figure 5.9: Histograms of proper motions in l and b of all proper motion stars. Error bars mark the Poisson errors. Dotted lines represent the respective best-fitting Gaussians. Average velocities $\langle v_{l,b} \rangle$ and velocity dispersions $\sigma_{l,b}$ are given in the plot. Whereas the b velocities appear to be normally distributed, the l velocities show a clear rotation pattern.

As shown above, the geometry of the cluster kinematics is in good agreement with the orientation of the Galaxy. Thus relative Galactic coordinates l, b are a more natural coordinate system than ecliptic coordinates α, δ. In the following discussion we will therefore preferentially focus on coordinates and velocities transformed into relative Galactic coordinates, using $(l, b)_{\text{SgrA}*} = (0, 0)$.

In stellar line-of-sight velocities, rotation in the sense of general Galactic rotation was already reported earlier (McGinn et al. 1989; Genzel et al. 1996 [and references therein]). Therefore we computed for our ~660 radial velocity stars (which form a subset of the proper motion stars) the average stellar radial velocities in given l bins. The resulting pattern is shown in the bottom panel of Fig. 5.8. Using a linear fit to describe the data points, we find the velocities to be zero within the errors (4.8 km/s) at $l = 0$, to be positive (i.e. receding from the observer) towards positive Galactic longitudes, and to be negative (i.e. approaching the observer) towards negative l, as expected. Our fit corresponds to the model of a rigid rotator. This does however not imply that the GC cluster indeed is a solid rotator; given the limited accuracies of the data, using a more complex rotation model is not justified. Keeping this in mind, we find a rotation velocity of 1.42 ± 0.36 km/s/arcsec. This corresponds to a 4-σ-detection of the Galactic rotation in radial velocities for $|l| \leq 24$ arcsec.

5.5. Results and discussion

The velocity distributions in l and b for all $\sim$5500 proper motion stars are shown in Fig. 5.9 together with the respective best-fitting Gaussian profiles. The velocities in b appear to be normally distributed. In contrast, the histogram of the l velocities shows clear broadening and flattening. The pattern can be approximately described as a convolution of a Gaussian with width σ_b and two δ-peaks located at roughly ± 2.5 mas/yr. This corresponds to the edge-on view through a system rotating with a fixed rotation velocity of $\sim$2.5 mas/yr. However, this number is an averaged and projected value and must not be read as the physical rotation velocity.

5.5.2 Phase-space distributions

As shown before, the observed part of the GC star cluster can be described as an isotropic rotator with normally distributed random stellar velocities. It is, however, not a priori clear if the observed stars kinematically indeed form a single system. A well-known example for kinematic segregation in the GC is the dichotomy between isotropically distributed late-type stars and early-type stars arranged in disks (Genzel et al. 2003a; Paumard et al. 2006).

A common way to characterize a stellar system is the use of phase-space maps. A nice example is the recent analysis of the Galactic thin-disk age vs. velocity dispersion relation by Seabroke & Gilmore (2007). With the help of velocity-velocity diagrams they are able to safely identify the Hyades cluster and the Hercules stream in their tracer population. Another example is the recent discovery of Galactic star streams based on their peculiar phase-space distributions (e.g. Ibata et al. 2001; Yanny et al. 2003; Martinez-Delgado et al. 2004).

For each star we have two or three velocity components and two coordinates at hand. By pairwise comparison of the velocities $v_{l,b,z}$ and positions l, b, we can construct a variety of phase-space diagrams. The resulting distributions are presented in Figs. 5.10 and 5.11. Qualitatively, there appears to be no substructure or grouping. In the v_z-l-diagram the data are biased towards positive l. This is an observational artefact as SINFONI spectra were preferentially obtained north of SgrA*, roughly corresponding to $l > 0$.

The phase-space maps also mirror the influence of global rotation discussed in the previous subsections. On the one hand, the rotation shows up as a broadening of the v_l-vs-coordinate distributions along the velocity axes compared to the respective distributions for v_b. On the other hand, one can hardly recognize the rotation pattern from the v_z-l plot as the random scatter of the data points (i.e. the dispersion) is much larger than the modulation in the velocity average (cf. Fig. 5.8 [bottom panel]).

In order to quantify the presence (or absence) of phase-space substructure, we made

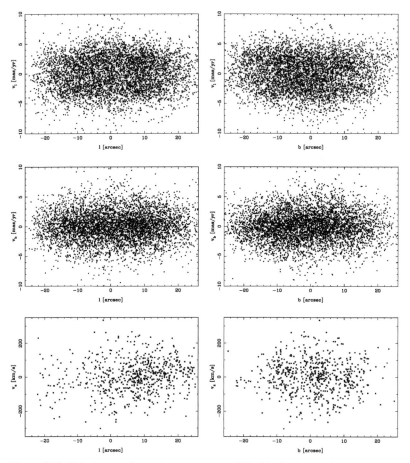

Figure 5.10: Velocity-coordinate phase-space maps. Velocities in the sky plane $v_{l,b}$ include all ∼5500 proper motion stars; line-of-sight velocities v_z are given for ∼660 stars. The scale is 1 mas/yr ≡ 37.9 km/s. Compared to the v_b, the v_l distributions are broadened (in velocity) due to the global rotation. The v_z are biased towards positive l as SINFONI spectra were collected mainly north of SgrA*. These global properties aside, the diagrams show no obvious patterns or sub-structures.

5.5. Results and discussion

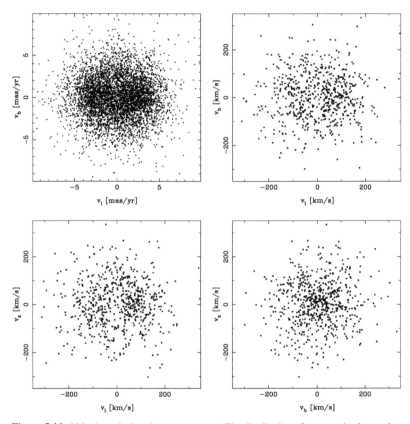

Figure 5.11: Velocity-velocity phase-space maps. The distribution of v_l vs. v_b is given twice, once for all ∼5500 proper motion stars (*top left*) in mas/yr, once for all ∼660 3D motion stars (*top right*) in km/s. The scale is 1 mas/yr ≡ 37.9 km/s. In analogy to Fig. 5.10, the global rotations shows up as a broadening of the $v_{l,z}$ distributions with respect to v_b and a slight density excess at $v_l \simeq \pm 100$ km/s (see also Figs. 5.9, 5.14). These global properties aside, there is no additional substructure or grouping.

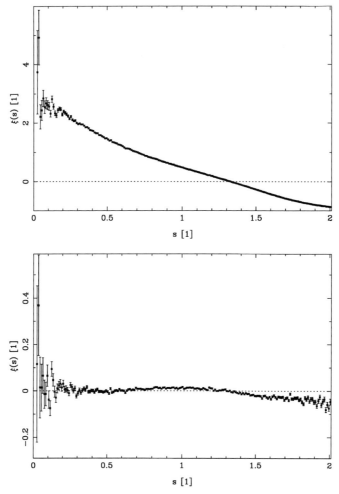

Figure 5.12: 4D two-point correlation functions for all proper motion stars. The normalized distance s is defined via $s^2 = (l/x_0)^2 + (b/x_0)^2 + (v_l/v_0)^2 + (v_b/v_0)^2$, with x_0, v_0 being constants. Please note the different $\xi(s)$ axis scales. *Left hand panel*: Measured phase-space distances compared to a uniform random distribution. The profile mirrors the obvious clustering of phase-space points around $(l, b, v_l, v_b) = (0, 0, 0, 0)$ shown in Figs. 5.10, 5.11. *Right hand panel*: Measured phase-space distances compared to an ad-hoc Monte-Carlo model of the GC cluster. Deviations from the uniform, isotropic rotator model cluster are smaller than 0.05 (0.02 for $s < 1.3$).

5.5. Results and discussion

use of the two-point correlation function (TPCF)

$$\xi(s) = \frac{n_R}{n_D} \cdot \frac{DD(s)}{DR(s)} - 1 \qquad (5.4)$$

(Davis & Peebles 1983). For a given distance s, $DD(s)$ is the number of pairwise distances between observed stars (also referred to as the data–data distances) located in the respective distance bin. $DR(s)$ is the number of pairwise distances between the observed stars and the members of a comparison ensemble, usually a random, uniform one (thus $DR(s)$ is also referred to as the data–random distances). $n_D(n_R)$ is the total number of data–data (data–random) distances. By definition, $\xi(s)$ is located in the range $[-1;+\infty]$; $\xi(s) = 0$ corresponds to full agreement between observed ensemble and comparison ensemble.

We computed the TPCF for all $\sim$5500 proper motion stars. As in phase-space the distance s mixes positions and velocities, we used a normalized distance

$$s = \sqrt{(l/x_0)^2 + (b/x_0)^2 + (v_l/v_0)^2 + (v_b/v_0)^2}$$

x_0, v_0 are (a priori arbitrary) constant distances and velocities. In order to match the typical phase-space dimensions of the cluster (cf. Figs. 5.10,5.11), we chose $x_0 = 1$ pc, $v_0 = 260$ km/s. Thus $s = 1$ corresponds to the half side length of a "phase space unit cell".

In a first step, we computed the TPCF using a uniform, random comparison ensemble. The resulting distribution is shown in Fig. 5.12 (left hand panel). It mirrors the obvious clustering of phase-space points around $(l,b,v_l,v_b) = (0,0,0,0)$ shown in Figs. 5.10,5.11. The point where the profile crosses the $\xi(s) = 0$ line can be identified as a characteristic phase-space radius of the cluster; this radius is $s_c \simeq 1.3$.

In a second step, we computed the TPCF using an ad-hoc Monte Carlo model of the cluster. This model is not a dynamical cluster model. It was constructed thus that the main observed kinematic properties are reproduced, namely the velocity dispersions in l,b and their modulation by rotation. It assumes a spherical, uniform distribution of stars following an isothermal sphere (e.g. Binney & Tremaine 1987) profile. The model data were extracted in two steps: (1) Computing a realization of the model cluster, and (2) application of a selection mask simulating the actual observation and the respective selection of stars (FOV, minimum star-star distances). These model data were inserted into the TPCF calculation.

The right hand panel of Fig. 5.12 shows the resulting $\xi(s)$ profile. Except for the largest s (above 1.5), deviations between observed cluster and uniform model are limited to less than 2% (in units of $\xi(s) + 1$). The largest deviations of about 5% are reached at $s \simeq 2$. The fact that systematic differences are present indicates that the choice of parameter values fed into the model cluster is not yet the optimum choice. Given the very small differences between observation and model, we can conclude the following: compared to a uniform, spherical, isotropic rotator, fluctuations or substructure are not present above a level of at most 5%. When limiting the comparison to $s < s_c$, this limit is even stricter: 2%. This leaves essentially no room for any type of phase-space clumping.

We did not include stellar radial velocities into this analysis. The spectra were extracted from 24 separate SINFONI data sets with different FOVs, pointings, pixel scales, integration times, spectral ranges, PSFs, and limiting magnitudes. This prevented a consistent reconstruction of the observation/selection mask, thus excluding the reliable extraction of a model cluster.

In total we can conclude that the GC star cluster appears as a uniform, well phase-mixed system. This is in good agreement with the age estimate of $\sim$12 Gyr (Maness et al. 2007), which is an order of magnitude beyond the relaxation time of $\sim 10^9$ yr (Alexander 2005). More interestingly, this also implies that the cluster did not experience a serious distortion within the last few 10^8 yr.

This finding nicely ties in with the long-standing debate on the origin of the young ($\sim$6 Myr) early-type stars in the GC, for which two mechanisms have been proposed: in-situ star formation (e.g. Levin & Beloborodov 2003; Goodman 2003) and immigration via an inspiraling star cluster (e.g. Gerhard 2001; McMillan & Portegies Zwart 2003). In their analysis, Paumard et al. (2006) conclude that the initial masses of infalling clusters is limited to $\sim$24,000 $M_\odot$ in total. In contrast to this, the inpiraling-cluster-scenario requires initial cluster masses $> 10^5$ $M_\odot$. From this, Paumard et al. (2006) conclude that the infall scenario is highly unlikely.

This conclusion is in good agreement with our findings. Given the relaxation time for the GC cluster ($\sim 10^9$ yr), kinematic structures (e.g. tidal tails) imprinted on the cluster few Myr ago by infalling objects should still be present. The fact that the phase-space maps do not show any corresponding feature implies that, if at all, only a small fraction of the GC stars could have been involved in or affected by such an event. Within the central parsec, the amount of stellar mass is however limited to roughly $2 \cdot 10^6$ $M_\odot$ (Genzel et al. 1996, 2000; Ghez et al. 1998). This makes it is hard to understand how a system containing about 10% of this mass could have entered the GC cluster without leaving behind substantial kinematic traces – but is in full agreement with the mass limits obtained by Paumard et al. (2006).

5.5.3 Distribution of stellar 3D speeds

For a dynamically relaxed stellar system, the stars' 3D speeds follow a Maxwellian distribution (e.g. von Hoerner 1960; Binney & Tremaine 1987). In case of the GC cluster, the discussion is somewhat complicated by the fact that it is dominated by a central point mass. As already observed by Genzel et al. (1996) and Ghez et al. (1998), the *local* velocity dispersion thus scales with the projected distance from SgrA*, r, like $\sigma \propto \sqrt{r}$. This means that the cluster is not isothermal. For a given tracer population of stars the 3D speeds should follow a Maxwellian distribution corresponding to the *average* ensemble velocity dispersion – if the GC cluster was actually dynamically relaxed.

In order to test the distribution of the GC star speeds, we analyzed the 664 stars with known 3D velocities. For each star we computed a bias-corrected 3D speed

$$v_{3D} = \sqrt{v_x^2 + v_y^2 + v_z^2 - \delta v_x^2 - \delta v_y^2 - \delta v_z^2}$$

where $v_{x,y,z}$ are the velocities (in km/s, assuming $R_0 = 8$ kpc) and $\delta v_{x,y,z}$ are the respective statistical errors.

In a separate step, we calculated the 3D ensemble velocity dispersion via

$$\sigma_{3D} = \sqrt{\sigma_x^2 + \sigma_y^2 + \sigma_z^2}$$

We found a value of $\sigma_{3D} = 179 \pm 5$ km/s. Using σ_{3D}, we computed a Maxwellian distribution for 664 stars. We would like to point out that σ_{3D} and the number of stars are the only observational constraints fed into the calculation. This means that the theoretical curve was obtained independently from the observed distribution – it is not a fit to the data.

Observed and theoretical distributions are compared in Fig. 5.13. A reduced-χ^2 test finds $\chi_{red}^2 = 0.92$, indicating a very good agreement. This shows that indeed the GC cluster stars follow a Maxwellian distribution.

In a stellar system following a Maxwellian distribution, the rms escape speed for a star is reached if $v_{3D}^2 > 4 \cdot \sigma_{3D}^2$ (i.e. $v_{3D} > 358$ km/s; e.g. Binney & Tremaine 1987). From integrating a Maxwellian, we expect this to be the case for five out of 664 stars; we actually observe 11. We even find three stars with $v_{3D}^2 > 6 \cdot \sigma_{3D}^2$ (i.e. $v_{3D} > 438$ km/s; from a Maxwellian distribution, only 0.3 are expected. Using a Monte Carlo test (with 10,000 realizations) operating on a Maxwellian distribution for 664 stars, we find the probability for three stars having speeds larger than 438 km/s to be 0.4%. This false alarm probability corresponds to a Gaussian significance of $\sim 2.8\sigma$.

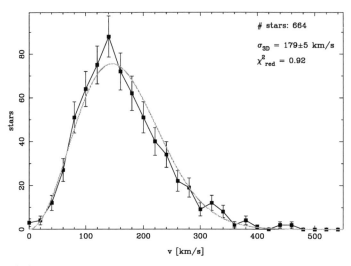

Figure 5.13: Distribution of stellar speeds for all 3D velocity stars. Points with error bars are data, error bars are Poisson errors. The continuous grey line corresponds to a Maxwellian distribution for a velocity dispersion $\sigma_{3D} = 179$ km/s; this line is *not* a fit to the data. Theoretical and observed distributions are in good agreement; we find $\chi^2_{red} = 0.92$.

Although of moderate significance, this excess might point towards non-Gaussian processes affecting these stars. In a recent analysis, Perets, Hopman & Alexander (2007) point out that the presence of so-called massive perturbers (mainly giant molecular clouds) in the GC region might lead to a substantial number of close binary star – SgrA* encounters. Those three-body interactions can result in binary disruption with one of the stars being ejected from the GC with a speed up to several thousand km/s (Hills 1988). This scenario is of interest especially in view of Galactic hypervelocity stars which are assumed to originate from the Galactic Center (Brown et al. 2005).

Table 5.1 summarizes the properties of the three stars with $v_{3D} > 438$ km/s. None of them shows a motion radially outward from SgrA*. More interesting, they all have projected distances from SgrA* of about 4" (0.15 pc), i.e. relatively close to the black hole. This is important as the local velocity dispersion increases with decreasing r. Using the r as lower limits for the physical distances from SgrA*, we obtain (using $M_\bullet = 4.1 \cdot 10^6 \, M_\odot$) for each star the maximum local escape speed

$$v_{esc,max} = \sqrt{\frac{2GM_\bullet}{r}}$$

From Table 5.1 one can conclude that the stars are not as special as their location in

5.5. Results and discussion

ID	l	b	v_l	v_b	v_z	v_{3D}	$v_{esc,max}$
407	+0.16	+3.36	+283	+41	+359	459	520
547	−3.62	−1.54	+147	+48	−432	459	480
627	+0.67	+4.11	+347	+268	+24	439	467

Table 5.1: Properties of the three stars with speeds $v_{3D} > 438$ km/s. Coordinates l,b are given in arcsec, velocities are given in km/s.

the global distribution suggests. In all three cases, the observed speeds are lower than the maximum escape speeds. Thus the stars are fast, but might still be bound. Although these objects somewhat stick out of the *global* distribution, they are not especially remarkable when taking into account the *local* conditions.

In total, we can conclude that (1) the Maxwellian distribution of the stellar 3D speeds confirms the relaxed nature of the GC late type cluster, and (2) that it is not necessary to assume non-Gaussian processes (like binary disruption) to understand the observations.

This result also helps to put the high speed of the SiO maser star IRS 9 (cf. Fig. 5.1) recently discussed by Reid et al. (2007) into a context. They found a 3D speed of ∼370 km/s for this star located 0.33 pc away from SgrA*. They concluded that IRS 9 is too fast to be bound to the mass enclosed within its distance from SgrA*. From our analysis (see Fig. 5.13) one can see that IRS 9 has a high, but not excessive 3D velocity with respect to the global speed distribution. As we find 11 out of 664 (i.e. 1.7%) stars with speeds above 358 km/s, detecting one out of 15 like in the Reid et al. (2007) sample does not appear exceptional.

5.5.4 Statistical parallax of the Galactic Center

The availability of 3-dimensional velocity vectors for several hundred stars allows the computation of the distance to the Galactic Center R_0 using the statistical parallax. If the stellar velocities are distributed isotropically, then the three velocity dispersions $\sigma_x, \sigma_y, \sigma_z$ are equal. As σ_x, σ_y are measured in angular units (mas/yr) whereas σ_z is measured in physical units (km/s), the distance scale can be derived immediately.

As discussed in section 5.1, the assumption of *global* isotropy is invalid for the GC star cluster due to the influence of the Galactic rotation. For the three velocity dispersions in l, b, and z of our ∼660 3D velocity stars we find the values

$$\sigma_l = 2.928 \pm 0.082 \text{ mas/yr}$$
$$\sigma_b = 2.531 \pm 0.071 \text{ mas/yr}$$
$$\sigma_z = 102.2 \pm 2.8 \text{ km/s}$$

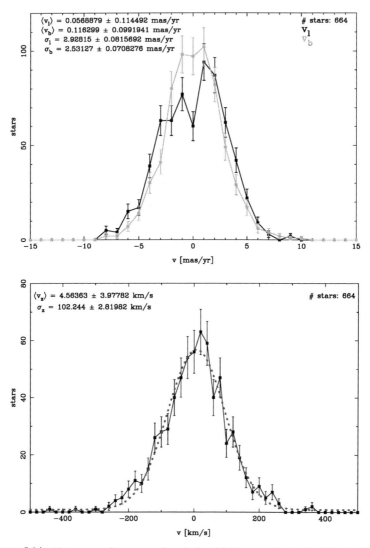

Figure 5.14: Histograms of proper motions in l and b (*top panel*) and of radial velocities (*bottom panel*) including all stars with measured 3D velocities. The dotted curve in the bottom panel corresponds to the best-fitting Gaussian profile. Error bars mark the Poisson errors. In both plots the respective average velocities $\langle v_{l,b,z} \rangle$ and velocity dispersions $\sigma_{l,b,z}$ are given.

5.5. Results and discussion

the respective distributions are shown in Fig. 5.14. Our value for σ_z is in excellent agreement with the value of 100.9±7.7 km/s found by Figer et al. (2003) who analyzed a smaller sample of 85 CO absorption line stars. The systematic difference between σ_l and σ_b is detectable on a $\sim$3.7-σ level and has to be taken into account explicitely. Fortunately, we can expect the observed part of the late-type star cluster to be *intrinsically* isotropic. Therefore we can calculate the statistical parallax after removing the effects of rotation.

First, we corrected σ_l by subtracting the contribution of rotation of 2.0 (mas/yr)^2 computed in sect. 5.1. The corrected value is

$$\sigma_l' = 2.564 \pm 0.082 \text{ mas/yr}$$

which agrees with σ_b within the errors as required for an isotropic sub-system.

In order to obtain a rotation-free value for σ_z, we took into account the rotation velocity estimate of $\sim$1.4 km/s/arcsec. We subtracted from each star's radial velocity the respective interpolated circular velocity and computed the velocity dispersion out of these corrected stellar velocities. The resulting value is

$$\sigma_z' = 101.0 \pm 2.8 \text{ km/s}$$

Using the average of σ_l' and σ_b, which is $\langle\sigma\rangle = 2.548 \pm 0.054$ mas/yr, and compares this with σ_z', one obtains a GC distance of

$$R_0 = 8.37 \pm 0.29 \text{ kpc}$$

Compared to the value of 7.62 ± 0.32 kpc derived by Eisenhauer et al. (2005a), the value we find here is larger by about 1.7σ. Although this is statistically not a significant deviation, it is important to compare our result to those obtained from the observations of Keplerian stellar orbits around SgrA* (e.g. Eisenhauer et al. 2005a, Lu et al. 2006). Indeed the most recent estimates for R_0 from orbit analyses unveiled systematic uncertainties which had not been taken into account so far (Gillessen et al. *in prep.*; A. Ghez, J. Lu *priv. comm.*).

In total, we can conclude the following: (1) Statistical parallax measurements are – along with experiments based on precision stellar photometry (e.g. Paczynski & Stanek 1998; McNamara et al. 2000; Nishiyama et al. 2006; Groenewegen, Udalski & Bono 2008) – a valuable independent measurement of R_0. (2) The agreement between

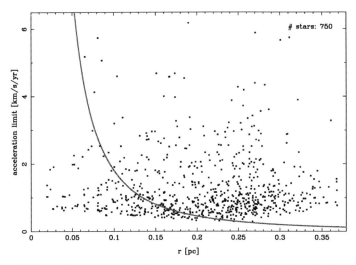

Figure 5.15: Acceleration upper limits vs. projected distance r from SgrA*. Dots mark measured values. The continuous line corresponds to the acceleration a star would experience if its physical distance from SgrA* would equal its projected distance. All values are given for $R_0 = 8$ kpc and $M_\bullet = 4.1 \cdot 10^6 \, M_\odot$. This sample includes 750 of 755 analysed stars. The large number of stars below the line indicates that typically the physical distance is much larger than the projected distance.

our value and measurements based on stellar orbits is better than suggested by a simple comparison of the statistical errors.

The value we find here is also somewhat larger than the results obtained by earlier statistical parallax measurements, which were 7.9 ± 0.9 kpc (Genzel et al. 2000) and, more recently, 7.1 ± 0.7 kpc (Eisenhauer et al. 2003a). This discrepancy is due to the influence of rotation. In the aforementioned experiments the systematic difference between the dispersions in l and b directions was masked by larger statistical uncertainties. Therefore the values (namely σ_l, $\langle\sigma\rangle$) used to derive the distance scale were systematically, but not obviously, too high. To give an illustration: if we compute $\langle\sigma\rangle$ out of the measured dispersions σ_l, σ_b instead of using the corrected value σ'_l, we find $\langle\sigma\rangle = 2.73$ mas/yr. This would lead (with σ_z instead of σ'_z) to $R_0 = 7.90$ kpc, i.e. to a value which is systematically too small by about 6% or 1.7σ, but which is more similar to those published so far.

5.5.5 Acceleration upper limits

The dominance of SgrA* allows the description the innermost part (few arcsec) of the cluster as a system of massless test particles moving around a point mass on Keplerian orbits. Indeed, orbits located in the innermost $\sim 0.5''$ – in the so-called 'S-star' group – have been observed now for several years without detecting any significant deviation from a point mass potential (Eisenhauer et al. 2005a; Lu et al. 2006; Gillessen et al. *in prep.*). When using the very accurate proper motions obtained from the small scale images (typical uncertainties ~ 4 km/s, see Fig. 5.4), it is possible to detect (or exclude) accelerations in stellar motion as far out as several arcseconds in projected distance.

In order to obtain acceleration limits we analysed the stars in the small scale fields within $\pm 7''$ in α, δ from SgrA*. All star positions were transformed into coordinates radial and tangential to their average position vector. In analogy to the determination of proper motions we fit the star positions q vs. time t as parabolas of the form

$$q = u \cdot t^2 + v \cdot t + w$$

Obviously, this approach delivers the (constant) stellar accelerations a (via $a = 2 \cdot u$); v and w correspond to velocities and positions at $t = 0$ respectively.

The decision of whether a given star shows a significant acceleration or not is based on the goodness-of-fit of the two physically realistic models, which are (1) a linear proper motion, and (2) an accelerated parabolic motion pointing towards SgrA* as described above. For both models the respective reduced χ^2 (hereafter χ^2_{lin} for the linear, χ^2_{acc} for the accelerated case) is computed.

In order to decide if the difference between the two models is significant we make use of the fact that the quantity

$$f = \frac{\chi^2_{\text{lin}}}{\chi^2_{\text{acc}}}$$

follows an F-distribution and can thus be examined using an F-test (e.g. Müller 1975; Lehn & Wegmann 1982). For a given significance level $S \in [0,1]$ the difference is considered to be significant if $f > F_{m,n,1-s/2}$; here m, n are the respective degrees of freedom, $s = 1 - S$ is the false alarm probability. For the typical case $m = n + 1 = 35$ we find $F_{35,34,0.995} = 2.45$ for $S = 0.99$. This means that a 99% confidence detection of non-linear motion requires $f > 2.45$.

Our analysis included a total of 755 stars *regardless of their spectral type*. For five of them significant (using a 99% confidence limit) accelerations were detected. Three stars were S-stars with known orbits. Two more were false positives whose centroids were systematically displaced towards bright neighbouring sources. One has to note

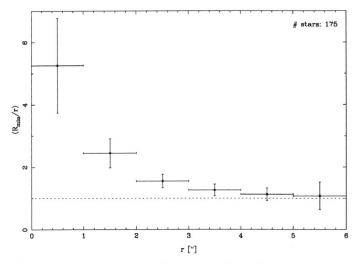

Figure 5.16: Ratio of minimum physical distance and projected distance from SgrA*, R_{min}/r as taken from acceleration upper limits, vs. projected distance. This sample includes all stars for which R_{min} exceeds r, the respective lower limit of 1 is indicated by the horizontal dotted line. Data points are the rms values for all stars in the respective bins. Horizontal error bars mark the full bin widths, vertical ones are standard errors of rms. This diagram shows the effect of projections: the more inward the selected subsample, the higher the minimum ratio between line-of-sight extension and plane-of-sky extension.

that due to the high crowding in the innermost arcseconds of the field most of the S-stars known to follow Keplerian orbits had to be excluded from this automatized analysis and therefore do not contribute.

For the remaining 750 sources not showing accelerations, acceleration upper limits are given. Acceleration limits δa are given as $\delta a = \delta v_{rad}/\Delta t$; δv_{rad} is the $1 - \sigma$ error of the proper motion in radial[3] direction, Δt is the total time covered by observations (about 5 years). Fig. 5.15 shows the resulting limits vs. projected distances. For comparison, we show the acceleration a star would experience if its projected distance were equal to its physical distance. This calculation uses $R_0 = 8$ kpc and $M_{\bullet} = 4.1 \cdot 10^6\ M_\odot$. The fact that many limits fall below the theoretical line indicates that typically the physical distance is much larger than the projected distance.

In order to quantify this projection effect, we analysed the distribution of the ratio of minimum physical distance R_{min} and projected distance r. The results of this analysis are summarized in Fig. 5.16. This diagram shows the rms of ratios R_{min}/r vs. r,

---
[3]Here the term "radial" refers to the line SgrA*–star.

5.5. Results and discussion

i.e. it describes the minimum extension along the line of sight with respect to the field of view for the respective subsamples. As stars for which $R_{min} < r$ do not constrain the geometry of the system, only stars with $R_{min} \geq r$ were considered. Indeed the diagram shows ratio values as large as $\sim$5 for the innermost arcsecond, meaning that the observed subsample of stars is more extended by a factor of at least 5 along the line of sight compared to its projected extension.

These results are of interest especially with respect to the ongoing search for stellar orbits in the innermost arcseconds. Firstly, the minimum line-of-sight extensions allow the constraint of detection probabilities. Secondly, the acceleration limits and/or the minimum physical distances derived thereby allow the constraint of orientations of orbits and their eccentricities; this becomes crucial in cases where no sufficient radial velocity data are available. Both points, which are beyond the scope of this article, play an important role in current and future work.

5.5.6 The star group IRS13E

An object of special interest is the star group IRS13E, located 3" west and 1.5" south of SgrA*. This object consists of three bright (H$\sim$13) main components concentrated within a region of about 0.2" radius. They surround fainter objects which are probably blends of several point sources. Paumard et al. (2006) found a significant stellar density excess in the immediate vicinity (0.7") of the three main stars and identified the IRS13E group as a star cluster. Based on stability arguments with respect to the tidal field of SgrA*, the possibility that IRS13E hosts an intermediate-mass black hole was discussed already by Maillard et al. (2004).

In order to examine this scenario in more detail, we tested if stars within a radius of 0.7" are kinematically connected with the three central sources. We extracted proper motions for the three main components and additional 17 stars with magnitudes down to H$\sim$19.5 (for a similar analysis using the proper motions of another set of stars, see Schödel, Eckart & Iserlohe 2005). In the following, we call the three main stars "set A" and the 17 field stars "set B". As the target area is too crowded to be fully covered by the automatic procedures described in section 3, we extracted image positions manually. We applied the PSF fitting routine *StarFinder* by Diolaiti et al. (2000) and checked the inter-epoch source identifications by eye. Our analyis included seven very good H and K-band small scale images obtained between 2002 and 2007.

The main results are summarized in Fig. 5.17. The top panel shows all stellar proper motions with respect to the standard astrometric reference frame tied to the GC cluster, i.e. co-moving with the GC cluster (see section 3). The bottom panel of this figure shows the same proper motions in a reference frame co-moving with set

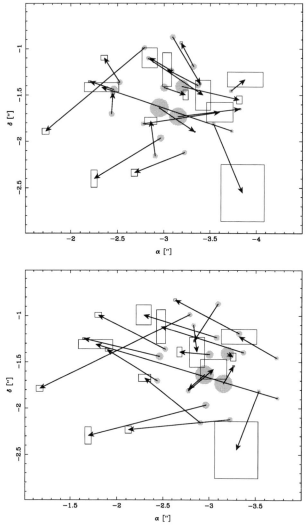

Figure 5.17: Proper motion maps for the IRS13E group. Filled circles are stars, circle diameters are proportional to the 0.25th power of H-band fluxes. Coordinates are α, δ in arcsec relative to SgrA*; please note the shift in α. Arrows are proper motions, 1 unit in length corresponds to 10 mas/yr. Boxes around arrow heads indicate the 1-σ velocity uncertainties. *Top panel*: Proper motions co-moving with the GC cluster as defined in section 3. *Bottom panel*: Proper motions co-moving with the central three brightest stars. A streaming motion of the three brightest stars relative to a foreground/background population can be seen. Apparently, if at all only a small fraction of the examined stars are bound members of the IRS13E group.

5.5. Results and discussion

A, using the average motion of set A as zero-point. In the later case, almost all stars show a motion directed from west to east, as if IRS13E was moving through a separate, non-co-moving foreground/background population. In α (or x) direction, the average motion (co-moving with set A) of set B is $\langle v_x \rangle = 7.1 \pm 1.3$ mas/yr, wheras the velocity dispersion is $\sigma_x = 5.3 \pm 0.9$ mas/yr. In δ (or y), we find $\langle v_y \rangle = 0.6 \pm 0.9$ mas/yr, $\sigma_y = 3.7 \pm 0.7$ mas/yr. These numbers show a clear streaming of the three main stars with respect to the 17 field stars.

For their analysis, Paumard et al. (2006 [see especially their Fig. 9]) used stellar number counts in the IRS13E region down to magnitudes H<20.4. They compared the surface densities inside ($\Sigma_<$) and outside ($\Sigma_>$) a projected radius of $p = 0.68"$ from the center of set A. Counting all H<20.4 stars they found

$\Sigma_< = 31.7 \pm 4.7$ arcsec^{-2}
$\Sigma_> = 13.1 \pm 1.2$ arcsec^{-2}

This corresponds to a 4.3σ excess of stars within p. Using a more conservative limit of H<19.4, the resulting densities are

$\Sigma_< = 17.9 \pm 3.5$ arcsec^{-2}
$\Sigma_> = 7.9 \pm 0.9$ arcsec^{-2}

corresponding to a 2.8σ excess of stars within p. All errors given for numbers and densities are Poisson errors. The total number of H<19.4 stars located within p is 26. This magnitude cut corresponds (within the errors) to the magnitude limit of our set B.

We used the kinematic information obtained for the set B stars to follow up on the surface density analysis by Paumard et al. (2006). We recalculated the surface density of H<19.4 stars, but excluded stars which are too fast with respect to set A. The star selection was done in two steps. In step one, we computed the 2D velocity dispersion of the three set A stars. This dispersion was $\sigma_A = 1.9 \pm 0.8$ mas/yr.

In step two, we computed the bias-corrected 2D speeds v_{2D} of the set B stars relative to set A (reference frame co-moving with set A). We identified all set B stars (a) located within p and (b) showing $v_{2D} > 3 \cdot \sigma_A$. For these stars we assumed that they cannot be physically connected to the set A stars.

We found nine stars satisfying criteria (a) and (b). Excluding them from the sample of Paumard et al. (2006) reduces the number of H<19.4 stars located within p from 26 to 17. Recalculating the surface density leads to

$$\Sigma_< = 11.7 \pm 2.8 \text{ arcsec}^{-2}$$

The difference of densities inside and outside p then becomes

$$\Delta\Sigma = \Sigma_< - \Sigma_> = 3.8 \pm 3.0 \text{ arcsec}^{-2}$$

meaning a significance of 1.3σ for a deviation from zero. We therefore do not see a significant excess of H<19.4 stars within p.

All in all, we can conclude that our kinematic analysis seriously weakens the scenario proposing that IRS13E is the core of a substantial star cluster.

5.6 Conclusions

In this article we analyzed and discussed the kinematic properties of the Galactic Center CO absorption line star cluster. This work is based on adaptive optics assisted diffraction-limited near-infrared imaging and integral-field spectroscopy. We collected proper motions for $\sim$5500 stars, 3D velocities for $\sim$660 stars, and acceleration limits for $\sim$750 stars. Our analysis led to the following main results:

1. The cluster shows a global rotation in the sense of general Galactic rotation.

2. The two-point correlation function of the stellar 4D phase-space positions agrees with that of an isotropic rotator within 2%. We find no indication for phase-space substructure, namely star streams. As the relaxation time is $\sim 10^9$ yrs, the infall of a $>10^5\ M_\odot$ star cluster into the central parsec within the last few Myrs is highly unlikely. This is in good agreement with the findings by Paumard et al. (2006).

3. The stellar 3D speeds follow a Maxwellian distribution. This confirms the relaxed nature of the CO star cluster. We do not find stars whose kinematics would require non-Gaussian processes like star ejection by three-body interactions.

4. Using the 3D velocity dispersion, we derive an improved statistical parallax to the GC of $R_0 = 8.37 \pm 0.29$ kpc. This result is slightly larger (on the $\sim 1.7\sigma$ level) than earlier ones as we for the first time take into account cluster rotation.

5. Upper limits on accelerations constrain the minimum line-of-sight extensions of observed star ensembles within the innermost few (projected) arcsec. The minimum ratio of true vs. projected extension reaches $\sim$5 for the innermost arcsec.

5.6. Conclusions

6. The star group IRS13E does not co-move with the $H < 19.4$ stars in its 0.7" vicinity. When ignoring stars which are too fast to be part of the IRS13E system, there is no sign for a significant star concentration. This seriously reduces the possibility that IRS13E is the core of a substantial star cluster.

In total, our analysis gained a substantial amount of knowledge regarding the kinematic properties of the GC cluster. The next step will be to feed our extensive data set into a full-scale dynamical model. We plan to make use of the recently developed *NMAGIC* code (De Lorenzi et al. 2007) in order to finalize the physical description of the CO star system in the central parsec of our Milky Way.

Acknowledgements: Special thanks to Nancy Ageorges (ESO, MPE) for helpful discussions on the instrumental geometric distortion and registration of NACO images. We are grateful to Tal Alexander and Hagai Perets (both: Weizmann Institute of Science, Rehovot, Israel) for enlightening discussions on non-Gaussian dynamical processes. S.T. would like to thank Markus Schmaus (LMU Munich, Dept. of Mathematics) for pointing out the concept of stochastic optimization.

Chapter 6

Conclusions

In this dissertation I described our studies of the center of the Milky Way. We addressed the following science cases:

1. The physical mechanism behind the photometric variability of the Ofpe/WN9 star IRS 34W.

2. Linearly polarized emission from Sagittarius A*, structure and dynamics of the emission region.

3. Kinematics of SiO maser stars and the alignment of radio and near-infrared coordinate systems.

4. Structure and kinematics of the CO absorption line star cluster.

In order to understand the physics behind the irregular photometric variability of IRS 34W, we analyzed data sets covering a time baseline of about 13 years. We made use of diffraction-limited H and K band imaging data obtained with the NIR cameras SHARP I and NAOS/CONICA. Additionally, we analyzed three K-band spectra obtained with 3D (in 1996), SPIFFI (in 2002), and SINFONI (in 2003). We found that (1) IRS 34W's flux is variable on timescales from months to years with amplitudes up to $\Delta K \sim 1.5$; (2) stellar colour and flux are correlated (the lower the flux, the redder the colour); (3) IRS 34W is significantly redder (by about 1 mag in K−L) and fainter than other Ofpe/WN9 stars in the field; (4) the (normalized) stellar spectrum is constant with time in contrast to the photometric variability, meaning that temperature and mass loss rate do not vary significantly.

This behaviour is best explained by variations in the column density of circumstellar material expelled by the star. The key point of this scenario is the lower intrinsic luminosity of IRS 34W compared to other Ofpe/WN9 stars. This allows the star to experience episodes of dust formation in its atmosphere. In each episode, the star gets redder and fainter. At the end of a dust formation period, the dust is partially destroyed by photo-dissociation. As this decomposition is incomplete, a long-term accumulation of dust occurs, making IRS 34W permanently redder and (even) fainter than other GC He I stars.

In order to understand better the physics behind the infrared emission from SgrA*, we have obtained regular photometric and polarimetric monitoring of the central black hole. On May 31st, 2006, we observed an exceptionally bright (up to $\sim$16 mJy) K-band outburst with NAOS/CONICA in polarimetry mode. Our observations covered the decreasing part of the event. From the lightcurves, we found the following properties of the flare: (1) The emitted light was highly polarized (polarization degrees 15-40%); (2) both the integrated and polarized flux curves show a double peak structure with a separation of $\sim$15 min between the maxima; (3) the polarization angle swings by about 70° within 15 min at the end of the outburst.

In the context of observations obtained earlier from radio to X-ray bands and various model calculations, we were able to draw the following picture: SgrA* is fairly rapidly (Kerr parameter $a > 0.6$) rotating. It is surrounded by an accretion disk with its inner edge close to the innermost stable circular orbit (ISCO). In an occasional (few times per day) magnetic reconnection event, a hot (electron temperature up to $\sim 10^{12}$ K), compact (size $< 0.3R_S$) plasma bubble arises at or close to the ISCO and orbits the black hole. The orbital period is of the order 15-20 minutes. The plasma electrons move in a toroidal magnetic field, emitting synchrotron emission. While orbiting SgrA*, the bubble is sheared and cools down. After about one to two hours, the cooled electron population moves further inwards into a region with a poloidal magnetic field – maybe a jet – and dissolves.

In order to improve the stability of astrometric reference frames for the Galactic Center field, we executed a dedicated combined NIR/radio monitoring program of nine SiO maser stars located in the central parsec. As SgrA* is a bright radio source, we could obtain time-resolved radio positions with respect to SgrA*. This defines a coordinate system tied to the central black hole. In parallel, we extracted time-resolved positions for the same stars from H and K band images. The NIR positions were tied to a reference ensemble of $\sim$500 stars, meaning a calibration relative to the infrared cluster. We computed the average motion of the stars from the radio and NIR data separately. By computing the difference of these two vectors we obtained the relative motion of the two coordinate systems.

As we used more stars and a longer timeline than earlier studies, we expected to find a reduced velocity difference due to reduced statistical uncertainties. Surprisingly, the relative motion turned out to be 30 ± 10 km/s. From this we had to conclude that our coordinate systems suffer from systematic uncertainties not understood so far.

Additionally, we found one star, IRS 9, to exceed (with 370 km/s at a projected distance from SgrA* of 0.33 pc) the local escape velocity. This behaviour is explained best by assuming that IRS 9 is indeed unbound *to the central $\sim$0.4 pc*. Instead, this star could move on a highly eccentric, "plunging" orbit. As the enclosed mass is increasing with radius, this would mean that IRS 9 could exceed the local escape speed, but still be bound to the cluster within few parsecs.

In order to gain a large-scale view on the stellar dynamics, we analyzed the kinematics of the CO absorption line star cluster within 1 pc from SgrA*. We extracted proper motions for about 5500 stars and line-of-sight velocities for about 660 stars. The accuracies reached unprecedented values of about 5-10 km/s per coordinate.

Both proper motion vectors and radial velocities show a pattern corresponding to global cluster rotation in the sense of the rotation of the Milky Way. From the 3D velocity dispersion we obtained a statistical parallax for the GC of 8.37 ± 0.29 kpc.

For about 750 stars we were able to derive acceleration limits in the plane of sky. This allowed us to constrain the minimum line-of-sight extension of several sub-ensembles of stars centered on SgrA*; for the innermost arcsecond, the minimum ratio of true vs. projected extension is about 5.

The stellar 3D speeds follow a Maxwellian distribution. This confirms the relaxed nature of the system. We do not find stars whose kinematics would require non-Gaussian processes (like 3-body-interactions with SgrA*).

We also were able to check the star cluster hypothesis for the IRS 13E star group. We compared the proper motions of the three main IRS 13E and a group of neighbouring (closer than $\sim$0.7") of surrounding stars. This comparison revealed that IRS 13E is moving separately from its neighbours. When ignoring stars which are too fast to be part of the IRS 13E system, there is no sign for a significant star concentration. These findings seriously reduce the possibility that IRS 13E and the surrounding stars form a substantial cluster.

The phase-space cuts of the cluster mirror the global properties mentioned above, especially the rotation. The two-point correlation function of the stellar 4D phase-space positions agrees with that of an isotropic rotator within 2%. We find no indication for phase-space subtructure, namely star streams. This shows that the cluster did not experience serious distortions, like the infall of another star cluster, within the last

$\sim 10^8$ years.

In total, the GC late-type star cluster is well described as a uniform, dynamically relaxed, phase-mixed system.

Since its early days, Galactic Center research in the infrared regime has been driven by technology. It began with the first seeing-limited NIR observations in the 1960s and cumulated in the emergence of adaptive optics assisted imagers and spectrographs at 8–10 m class telescopes after the year 2000. Each step forward in technology led to new insights into the physics of the Galactic Center. This thesis gives an overview of science cases that we were able to address with these state-of-the-art techniques.

With the highest probability, the next important step will be the implementation of near-infrared interferometry using baselines of the order $\sim$100 m. This will lead to improvements in angular resolution by factors about 10. One instrument currently under development by an MPE-led consortium is GRAVITY, an interferometer for the VLT (Eisenhauer et al. 2005b). When fully operational, GRAVITY will combine the light from the four 8.2-m VLT Unit Telescopes. This will allow the attainment of astrometric accuracies corresponding to the angle of the Schwarzschild radius of SgrA* on sky. Exciting new discoveries can be expected.

Acknowledgements

For this dissertation the collaboration with and support by a number of persons was decisive.

At first I am grateful to my supervisor Prof. Reinhard Genzel for the opportunity to work in a well-recognized, high-level research group offering a great scientific spirit, and also for his advise and encouragement.

I would like to thank my secondary supervisor Dr. Thomas Ott for introducing me into the exciting field of Galactic Center research and permanently providing scientific and technical assistance.

Many thanks also ...

To all the current and past members of the Galactic Center team for the intense collaboration and common work during the last four years: Frank Eisenhauer, Fabrice Martins, Stefan Gillessen, Roberto Abuter, Katie Dodds-Eden, Sonja Rank, Holly Maness, Nico Hamaus, Hendrik Bartko, and Thibaut Paumard.

To Mario Schweitzer, Francisco Müller-Sanchez, Peter Buschkamp, Giovanni Cresci, Yohei Harayama, and Sebastian Ihle for their friendship and many joyful hours.

To those doing most of the "backstage" management, namely Dr. Linda Tacconi and Susanne Harai-Ströbl.

To Sebastian Rabien, Ric Davies, Dieter Lutz, Eckart Sturm, Albrecht Poglitsch, Thomas Müller, and all the other members of the infrared group sharing their expertise and offering support.

To my father, for all his love, assistance, and encouragement.

To my mother, in memoriam.

To my siblings, Daniel and Rebecca, for being there whenever necessary.

To Maria, "Ibo" Ibrahim, and the "Wesel branch" of the Marciszyn family – Lukasz, Klaudia, Dawid, and Emely – for their friendship and curiosity.

Consummatum est.

Bibliography

[1] Abuter, R., et al. 2006, NewAR, 50, 398

[2] Aitken, D.K., et al. 2000, ApJ, 5334, L173

[3] Alexander, T. 2005, PhR, 419, 65

[4] Alexander, T. & Loeb, A. 2001, ApJ, 551, 223

[5] Aschenbach, B., et al. 2004, A&A, 417, 71

[6] Aschenbach, B. 2006, ChJAS, 6, 221

[7] Backer, D. C. & Sramek, R. A. 1999, ApJ, 524, 805

[8] Baganoff, F.K., et al. 2001, Nature, 413, 45

[9] Baganoff, F.K., et al. 2003, ApJ, 591, 891

[10] Bahcall, J.N. & Wolf, R.A. 1976, ApJ, 209, 214

[11] Bahcall, J.N. & Tremaine, S. 1980, ApJ, 244, 805

[12] Balick, B. & Brown, R.L. 1974, ApJ, 194, 265

[13] Bardeen, J.M., Press, W.H. & Teukolsky, S.A. 1974, ApJ, 194, 265

[14] Bates, J.H.T., Fright, W.R. & Bates, R.H.T. 1984, MNRAS, 211, 1

[15] Baumgardt, H., Makino, J. & Hut, P. 2005, ApJ, 620, 238

[16] Becklin, E.E. & Neugebauer, G. 1968, ApJ, 151, 145

[17] Becklin, E.E. & Neugebauer, G. 1975, ApJ, 200, L71

[18] Becklin, E.E.Gatley, I. & Werner, M.W. 1982, ApJ, 258, 135

[19] Bélanger, G., et al. 2005, ApJ, 635, 1095

[20] Bélanger, G., et al. 2006, JPhCS, 54, 420

[21] Beloborodov, A. M., et al. 2006, ApJ, 648, 405

[22] Bender, R., et al. 2005, ApJ, 631, 280

[23] Berry, R. & Burnell, J. 2000, *Handbook of astronomical image processing*, Willmann-Bel, ISBN 0-94339-667-0

[24] Bianchi, L., et al. 2004, ApJ, 601, 228

[25] Binney, J. & Tremaine, S. 1987, *Galactic Dynamics*, Princeton University Press

[26] Blum, R.D., Sellgren, K. & DePoy, D.L. 1996, ApJ, 470, 864

[27] Blum, R.D., Sellgren, K. & DePoy, D.L. 1996, AJ, 112, 1988

[28] Bohannan, B. & Walborn, N.A. 1989, PASP, 101, 639

[29] Bohannan, B. & Crowther, P.A. 1999, ApJ, 511, 374

[30] Bonnet, H., et al. 2003, SPIE, 4839, 329

[31] Bonnet, H., et al. 2004, The ESO Messenger, 117, 17

[32] Bourdet, G.L., et al. 1978, Proceedings of the Conference on Optical telescopes of the Future (Geneva), 445

[33] Bower, G. C. & Backer, D. C. 1998, ApJL, 496, L97

[34] Bower, G.C., et al. 1999, ApJ, 521, 582

[35] Bower, G.C., Falcke, H. & Backer, D.C. 1999, ApJ, 523, L29

[36] Bower, G.C., et al. 2003, ApJ, 588, 331

[37] Bower, G.C. 2003, Ap&SS, 288, 69

[38] Bower, G. C., et al. 2004, Science, 304, 704

[39] Bower, G.C., et al. 2005, ApJ, 618, L29

[40] Brandl, B. 1996, PhD thesis, Ludwig-Maximilians-Universität München

[41] Bresolin, F., et al. 2002, ApJ, 577, L110

[42] Broderick, A.E. & Loeb, A. 2006, MNRAS, 367, 905

[43] Bronstein, I.N., et al. 1999, *Taschenbuch der Mathematik*, 4th ed., Verlag Harri Deutsch

[44] Brown, W. R., et al. 2005, ApJL, 622, L33

[45] Carroll, B.W. & Ostlie, D.A. 1996, *An Introduction to Modern Astrophysics*, Addison-Wesley, ISBN 0-201-54730-9

[46] Chatterjee, P., Hernquist, L., & Loeb, A. 2002, ApJ, 572, 371

[47] Cherchneff, I. & Tielens, A.G.G.M. 1995, IAUS, 163, 346

[48] Christou, J.C. 1991, Exp. Astron., 2, 27

[49] Clark, J.S., et al. 2003a, A&A, 403, 653

[50] Clark, J.S., et al. 2003b, A&A, 412, 185

[51] Clark, J.S., et al. 2005, A&A, 435, 239

[52] Clénet, Y., et al. 2001, A&A, 376, 124

[53] Clénet, Y., et al. 2004, A&A, 424, L21

[54] Clénet, Y., et al. 2005, A&A, 439, L9

[55] Conti, P.S. 1984, IAUS, 105, 233

[56] Crowther, P.A., et al. 1995, A&A, 293, 172

[57] Crowther, P.A., et al., 2002, ApJ, 579, 774

[58] Davis, M. & Peebles, P.J.E. 1983, ApJ, 267, 465

[59] Dehnen, W. & Binney, J. 1998, MNRAS, 294, 429

[60] De Lorenzi, F., et al. 2007, MNRAS, 376, 71

[61] De Villiers, J.-P., Hawley, J.F. & Krolik, J.H. 2003, ApJ, 599, 1238

[62] Diamond, P. J. & Kemball, A. J. 2003, AJ, 599, 1372

[63] Diolaiti, E., et al. 2000, A&AS, 147, 335

[64] Doeleman, S. S., et al. 2001, AJ, 121, 2610

[65] Dorfi, E.A. & Gautschy, A. 2000, ApJ, 545, 982

[66] Downes, D. & Maxwell, A. 1966, ApJ, 146, 653

[67] Eckart, A., et al. 1995, ApJ, 445, L23

[68] Eckart, A. & Genzel, R. 1996, Nature, 383, 415

[69] Eckart, A. & Genzel, R. 1997, MNRAS, 284, 576

[70] Eckart, A., et al. 2002, MNRAS, 331, 917

[71] Eckart, A., et al. 2006a, A&A, 450, 535

[72] Eckart, A., et al. 2006b, A&A, 455, 1

[73] Eisenhauer, F., et al. 2003, ApJ, 597, L121

[74] Eisenhauer, F., et al. 2003, SPIE, 4814, 1548

[75] Eisenhauer, F., et al. 2003, The ESO Messenger, 113, 17

[76] Eisenhauer, F., et al. 2005, ApJ, 628, 246

[77] Eisenhauer, F., et al. 2005, AN, 326, 561

[78] Ekers, R.D., et al. 1983, A&A, 122, 143

[79] Falcke, H. & Markoff, S. 2000, A&A, 362, 113

[80] Ferrarese, L. & Ford, H. 2005, SSRv, 116, 523

[81] Figer, D.F., McLean, I.S. & Morris, M. 1999, ApJ, 514, 202

[82] Figer, D.F., et al. 2003, ApJ, 599, 1139

[83] Fomalont, E.B., et al. 1992, MNRAS, 258, 497

[84] Freitag, M., Amaro-Seoane, P. & Kalogera, V. 2006, ApJ, 649, 91

[85] Genzel, R., et al. 1996, ApJ, 472, 153

[86] Genzel, R., et al. 1997, MNRAS, 291, 219

[87] Genzel, R., et al. 2000, MNRAS, 317, 348

[88] Genzel, R., et al. 2003, ApJ, 594, 812

[89] Genzel R., et al. 2003, Nature, 425, 934

[90] Gerhard, O. 2001, ApJ, 546, L39

[91] Ghez, A.M., et al. 1998, ApJ, 509, 678

[92] Ghez, A.M., et al. 2000, Nature, 407, 349

[93] Ghez, A.M., et al. 2003, ApJ, 586, L127

[94] Ghez, A.M., et al. 2004, ApJ, 601, L159

[95] Ghez, A.M., et al. 2005, ApJ, 620, 744

[96] Ghez, A.M., et al. 2005, ApJ, 635, 1087

[97] Gillessen, S., et al. 2006, ApJ, 640, 163

[98] Glatzel, W. & Kiriakidis, M. 1993, MNRAS, 263, 375

[99] Goldwurm, A., et al. 2003, ApJ, 584, 751

[100] Goodman, J. 2003, MNRAS, 339, 937

[101] Groenewegen, M.A.T., Udalski, A. & Bono, G. 2008, A&A, in press

[102] Gustafsson, B., et al. 2003, ASP Conf. Ser. 288, 331

[103] Hargreaves, J.C., et al. 1994, MNRAS, 269, 957

[104] Hartung, M., et al. 2003, A&A, 399, 385

[105] Hartung, M., et al. 2003, SPIE, 4841, 425

[106] Heisler, J., Tremaine, S. & Bahcall, J.N. 1985, ApJ, 298, 8

[107] Hills, J. G. 1988, Nature, 331, 687

[108] Hofmann, R., et al. 1992, Proceedings of the ESO Conference on Progress in Telescope and Instrumentation Technologies, Garching, Germany, 617

[109] Högbom, J.A. 1974, A&AS, 15, 417

[110] Hornstein, S.D., et al. 2002, ApJL, 577, L9

[111] Hornstein, S.D., et al. 2007, ApJ, 667, 900

[112] Humphreys, R.M. & Davidson, K. 1984, Science, 223, 243

[113] Humphreys, R.M. & Davidson, K. 1994, PASP, 106, 1025

[114] Ibata, R.A., et al. 2001, ApJ, 547, L133

[115] Jähne, B. 2005, *Digital Image Processing*, Springer Verlag, Heidelberg

[116] Kassim, N.E., et al. 1999, ASP Conf. Series, 186, 403

[117] Kato, S. 2001, PASJ, 53, 1

[118] Kirkpatrick, S., Gelatt, C.D. & Vecchi, M.P. 1983, Science, 220, 4598

[119] Kormendy, J. & Bender, R. 1999, ApJ, 522, 772

[120] Krabbe, A., et al. 1991, ApJ, 382, L19

[121] Krabbe, A., et al. 1995, ApJ, 447, L95

[122] Krabbe, A., et al. 2006, ApJ, 642, L145

[123] Krichbaum, T.P. et al. 1998, A&A, 335, L106

[124] Labeyrie, A. 1970, A&A, 6, 85

[125] Lada, C.J. & Reid, M.J. 1978, ApJ, 219, 95

[126] Lafler, J. & Kinman, T.D. 1965, ApJS, 11, 216

[127] van Langevelde, H. J., et al. 1992, A&A, 261, L17

[128] Lamers, H.J.G.L.M. & Fitzpatrick, E.L. 1988, ApJ, 324, 279

[129] Lehn, J. & Wegmann, H. 1982, *Einführung in die Statistik*, Wissenschaftliche Buchgesellschaft, Darmstadt

[130] Léna, P., Lebrun, F. & Mignard, F. 1998, *Observational Astrophysics*, Springer, ISBN 3-540-63482-7

[131] Lenzen, R., et al. 2003, SPIE, 4841, 944

[132] Leonard, P.J.T. & Merrit, D. 1989, ApJ, 339, 195

[133] Levin, Y. & Beloborodov, A.M. 2003, ApJ, 590, L33

[134] Liu, S., Petrosian V. & Melia F. 2004, ApJ, L101, 2004

[135] Lu, J. R., et al. 2006, JPhCS, 54, 279

[136] Lucy, L.B. 1974, AJ, 79, No. 6, 745

[137] Lynden-Bell, D. & Rees, M. J. 1971, MNRAS, 152, 461

[138] Macquart, J.-P., et al. 2006, ApJ, 646, L111

[139] Maillard, J.P., et al. 2004, A&A, 423, 155

[140] Maness, H., et al. 2007, ApJ, 669, 1024

[141] Markoff, S., et al. 2001, A&A, 379, L13

[142] Marrone, D.P., et al. 2006, ApJ, 640, 308

[143] Martinez-Delgado, D., et al. 2004, ApJ, 601, 242

[144] Martins, F., et al. 2006, ApJ, 649, L103

[145] Massey, P. 2003, ARA&A, 41, 15

[146] McGinn, M. T., et al. 1989, ApJ, 338, 824

[147] McMillan, S.L.W. & Portegies Zwart, S.F. 2003, ApJ, 596, 314

[148] McNamara, D.H., et al. 2000, PASP, 112, 202

[149] Melia, F. & Falcke, H. 2001, ARA&A, 39, 309

[150] Menten, K.M., et al. 1997, ApJ, 475, L111

[151] Miralda-Escudé, J. & Gould, A. 2000, ApJ, 545, 847

[152] Miyazaki, A., Tsutsumi, T. & Tsuboi, M. 2004, ApJ, 611, L97

[153] Modigliani, A., et al. 2007, Proc. ADA IV, in press

[154] Moneti, A., et al. 2001, ApJS, 102, 383

[155] Morris, M. 1993, ApJ, 408, 496

[156] Morris, P.W., et al. 1996, ApJ, 470, 597

[157] Mouawad, N., et al. 2005, AN, 326, 83

[158] Müller, P.H. (ed.) 1975, *Lexikon der Wahrscheinlichkeitsrechnung und Mathematischen Statistik*, Akademie-Verlag, Berlin

[159] Najarro, F., et al. 1997, A&A, 325, 700

[160] Nayakshin, S., Cuadra, J. & Sunyaev, R. 2003, A&A, 413, 173

[161] Nishiyama, S., et al. 2006, ApJ, 647, 1093

[162] Nota, A., et al. 1996, ApJS, 102, 383

[163] Nugis, T. & Lamers, H.J.G.L.M. 2002, A&A, 389, 162

[164] Oort, J.H. & Rougoor, G.W. 1960, MNRAS, 121, 171

[165] Ott, T., Eckart, A. & Genzel, R. 1999, ApJ, 523, 248

[166] Pacholczyk, A.G. 1970, *Radio Astrophysics*, Freeman, ISBN 0-7167-0329-7

[167] Paczynski, B. & Stanek, C. 1998, ApJ, 494, L129

[168] Pasquali, A., et al. 1997, ApJ, 478, 340

[169] Paumard, T., et al. 2001, A&A, 366, 466

[170] Paumard, T., Maillard, J.-P. & Morris, M. 2004, A&A, 426, 81

[171] Paumard, T., et al. 2006, ApJ, 643, 1011

[172] Peebles, P.J.B. 1972, ApJ, 178, 371

[173] Perets, H.B., Hopman, C. & Alexander, T. 2007, ApJ, 656, 709

[174] Pétri, J. 2006, Ap&SS, 302, 117

[175] Rabien, S., et al. 2003, SPIE, 4839, 393

[176] Rank, S. 2007, MSc thesis, Technische Universität München

[177] Reid, M.J. 1993, ARA&A, 31, 345

[178] Reid, M.J. & Brunthaler, A. 2004, ApJ, 616, 872

[179] Reid, M.J. & Menten, K. M. 1997, ApJ, 476, 327

[180] Reid, M., et al. 2007, ApJ, 659, 378

[181] Reid, M.J., et al. 1999, ApJ, 524, 816

[182] Reid, M.J., et al. 2003, ApJ, 587, 208

[183] Repolust et al., 2004, A&A, 415, 349

[184] Richardson, W.H. 1972, J. Opt. Soc. Am., 62, 55

[185] Rieke, G. H. & Rieke, M. J. 1988, ApJ, 330, L33

[186] Roddier, F., Gilli, J.M. Lund, G. 1982, Journal of Optics, 13, 263

[187] Rogers, A. E. E., et al. 2004, ApJ, 434, L59

[188] Rousset, G., et al. 2003, SPIE, 4839, 140

[189] Salim, S. & Gould, A. 1999, ApJ, 523, 633

[190] Scargle, J.D. 1982, ApJ, 263, 835

[191] Schnittman, J.D., Krolik J.H. & Hawley J.F. 2006, ApJ, 651, 1031

[192] Schödel, R., et al. 2002, Nature, 419, 694

[193] Schödel, R., et al. 2003, ApJ, 596, 1015

[194] Schödel, R., Eckart, A. & Iserlohe, C. 2005, ApJ, 625, L111

[195] Schödel, R., et al. 2007, A&A, 469, 125

[196] Schuster, A. 1898, Terr. Magn., 3, 13

[197] Scoville, N.Z., et al. 2003, ApJ, 594, 294

[198] Seabroke, G.M. & Gilmore, G. 2007, MNRAS, 380, 1348

[199] Shen, Z.–Q., et al. 2005, Nature, 438, 62

[200] Sjouwerman, L.O., et al. 1998, A&A, 128, 35

[201] Smith, R.C. 1995, *Observational Astrophysics*, Cambridge Univ. Press, ISBN 0-521-26091-4

[202] Stahl, O., et al. 1983, A&A, 127, 49

[203] Stahl, O. 1986, A&A, 164, 321

[204] Stetson, P.B. 1987, PASP, 99, 191

[205] Stone, R.C. 1989, AJ, 97, 1227

[206] Timmer, J. & König, M. 1995, A&A, 300, 707

[207] Trippe, S. 2004, MSc thesis, Ludwig-Maximilians-Universität München

[208] Uttley, P., McHardy, I.M. & Papadakis, I.E. 2002, MNRAS, 332, 231

[209] Vaughan, S. 2005, A&A, 431, 391

[210] Viehmann, T., et al. 2005, A&A, 433, 117

[211] von Hoerner, S. 1960, ZA, 50, 184

[212] Voors, R.H.M., et al. 2000, A&A, 356, 501

[213] Wakker, B.P. & Schwarz, U.J. 1988, A&A, 200, 312

[214] Walborn, N.R. 1982, ApJ, 256, 452

[215] Wardle, M. & Yusef-Zadeh, F. 1992, ApJL, 387, L65

[216] Weitzel, L., et al. 1996, A&AS, 119, 531

[217] Wiener, N. 1950, *Extrapolation, interpolation, and smoothing of time series*, Wiley

[218] Winnberg, A, et al. 1985, ApJ, 291, L45

[219] Yanny, B., et al. 2003, ApJ, 588, 824

[220] Yu, Q. & Tremaine, S. 2003, ApJ, 599, 1129

[221] Yuan, F., Quataert, E. & Narayan, R. 2004, ApJ, 606, 894

[222] Yuan, F., Markoff, S. & Falcke, H. 2002, 383, 854

[223] Yusef-Zadeh, F., et al. 2006, ApJ, 644, 198

[224] Yusef-Zadeh, F., et al. 2006, ApJ, 650, 189

[225] Zoccali, M., et al. 2003, A&A, 399, 931

[226] Zylka, R. & Mezger, P.G. 1988, A&A, 190, L25

[227] Zylka, R., Mezger, P.G. & Lesch, H. 1992, A&A, 261, 119